PARTS

.01 Capacitor

1N4148 Diode

100K Potentiometer

555 Timer IC

1N4003 Diodes

TIP 3055 Transistors

T0260473

in its "astable mode" (basically as an oscillator) to send a bunch of short, spiky signals that the servo reads as the longer signal it's usually looking for. Attach a potentiometer and you can dial the servo to the position you want.

Stepper Controller

Our stepper controller also uses a clever circuit. While it's complicated relative to the other two circuits, it's much simpler than a microcontroller, which is usually required to control a stepper motor. Here again we employ a 555 chip and add two logic chips, a 4027 dual "flip-flop" and a 4070 XOR logic gate. The 555 generates the clock pulses and the two logic chips organize that signal into the correct sequence to drive the motor in "steps."

After completing these circuits, you'll have a far greater understanding of how each motor type works and what's required to control it. From there you can begin to design motor control for your robotic and other electronics projects. Get moving!

—Gareth Branwyn, MAKE editorial director,
and Steve Hobley

To see full build instructions, breadboard illustrations, and project photos, visit the project page for this build: radioshackdiy.com/project-gallery/projects-in-motion

Make: Volume 31

PUNK SCIENCE

ON THE COVER

HAVING A BALL: Finn the dog had an excellent workout as our Fetch-O-Matic tester. Photography by Gregory Hayes. Art direction by Jason Babler. Build by Daniel Spangler.

61

ROACH COACH: Have you ever dreamed of being able to remote-control a cockroach? Us, too!

70

FACING THE FUTURE: Syneseizure is a sight-to-touch mask built in 24 hours at Science Hack Day.

Vol. 31, July 2012. MAKE (ISSN 1556-2336) is published quarterly by O'Reilly Media, Inc. in the months of January, April, July, and October. O'Reilly Media is located at 1005 Gravenstein Hwy. North, Sebastopol, CA 95472, (707) 827-7000. SUBSCRIPTIONS: Send all subscription requests to MAKE, P.O. Box 17046, North Hollywood, CA 91615-9588 or subscribe online at makezine.com/offer or via phone at (866) 289-8847 (U.S. and Canada); all other countries call (818) 487-2037. Subscriptions are available for $34.95 for 1 year (4 quarterly issues) in the United States; in Canada: $39.95 USD; all other countries: $49.95 USD. Periodicals Postage Paid at Sebastopol, CA, and at additional mailing offices. POSTMASTER: Send address changes to MAKE, P.O. Box 17046, North Hollywood, CA 91615-9588. Canada Post Publications Mail Agreement Number 41129568. CANADA POSTMASTER: Send address changes to: O'Reilly Media, PO Box 456, Niagara Falls, ON L2E 6V2

Make: Projects

Sound-O-Light Speakers

Surprisingly simple, these clear PVC pipe speakers are shining performers.
By William Gurstelle

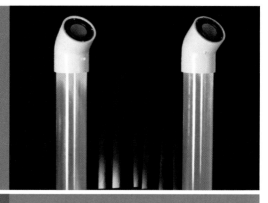

Rocket Glider

The classic toy, remade. It rockets up, then glides back down!
By Rick Schertle

Fetch-O-Matic

Build your own automatic tennis ball launcher for dogs.
By Dean Segovis

SERVO CONTROLLERS

Choose the best board for your build.
By Robert H. Walker

SKILL BUILDER

Brooke Davis blends traditional and digital tools to make one-of-a-kind objects of beauty.

WHERE Austin, TX

BUSINESS Brooke's original furniture & objects: *brookemdavisdesign.com*

Design consulting services & hackerspace: *makeshiftatx.com*

SHOPBOT PRSstandard 96 X 48

Tablescape No. 1

Brooke M Davis Design blurs the line between artisan and designer by combining artistic expression with design precision to produce luxury craftsmanship. Brooke's process of discovery and creation for Tablescape No.1 *(shown here)* is much like it is for all her original works.

"It's a fluid process, and it always involves my ShopBot CNC," notes Brooke. "I'm a hands-on person; I'd rather make an object in 3D than draw it. So I use digital fabrication to make physical models — you could call them 'interim prototypes' — and modify them throughout my process.

For instance, I'll add clay onto a model to get correct proportions visually (in furniture design). Or I'll use them to help me align measurements (like when creating a case to fit an iPad). Creating these interim prototypes is also great for client meetings. People respond to things they can touch and feel."

Austin Hackers, Take Note!

Brooke also founded the hacker space, make+SHift, in 2011. "It's the only Design on Demand Shop for product developers in Texas!" says Brooke. "We provide all the resources to make your ideas a reality. We offer design consulting services, CNC prototyping, and classes to advance professional development, plus a comfortable work space."

"One of the things that's been wonderful about opening make+SHift is how much I learn from my customers," says Brooke. "They bring new challenges to the table, and my team gets to think of new ways of solving problems. It energizes my own design thinking as well."

Allura SideTable

Make: Volume 31

READ ME: Always check the URL associated with a project before you get started. There may be important updates or corrections.

144

BACKYARD SPACE PROGRAM: This high-tech treehouse is loaded with fun features.

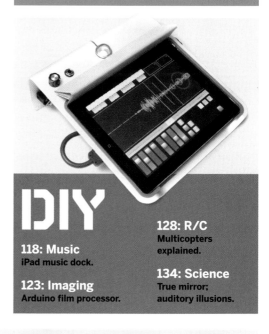

DIY

New To Making? Start Here!

This issue has a number of projects that are well-suited to new makers:

« Lord Kelvin's Thunderstorm (page 45) is a fun middle-school-level science fair project that creates electric sparks from falling water.

The levitating **Mendocino Motor** (page 64) is easy to build but will keep people guessing as to how it spins.

Our **Rocket Glider** (page 88) is available as a kit — build it after school and be launching it before sunset.

The broomstick **Monopod** (page 111) mounts to any camera with a standard ¼-20 threaded hole.

The **True Mirror** (page 134) flips your face and lets you see yourself the way others see you.

> "The true method of knowledge is experiment."
> —*William Blake*

Make: ®

FOUNDER & PUBLISHER
Dale Dougherty
dale@oreilly.com

EDITORIAL DIRECTOR
Gareth Branwyn
gareth@makezine.com

MAKER-IN-CHIEF
Sherry Huss
sherry@oreilly.com

EDITORIAL

EDITOR-IN-CHIEF
Mark Frauenfelder
markf@oreilly.com

PROJECTS EDITOR
Keith Hammond
khammond@oreilly.com

SENIOR EDITOR
Goli Mohammadi
goli@oreilly.com

ASSISTANT EDITOR
Laura Cochrane

STAFF EDITOR
Arwen O'Reilly Griffith

EDITORS AT LARGE
Phillip Torrone
David Pescovitz

WEBSITE
WEB PRODUCER
Jake Spurlock
jspurlock@oreilly.com

DESIGN & PHOTOGRAPHY

CREATIVE DIRECTOR
Jason Babler
jbabler@oreilly.com

SENIOR DESIGNER
Katie Wilson

SENIOR DESIGNER
Michael Silva

ASSOCIATE PHOTO EDITOR
Gregory Hayes
ghayes@oreilly.com

MAKER FAIRE

PRODUCER
Louise Glasgow

MARKETING & PR
Bridgette Vanderlaan

PROGRAM DIRECTOR
Sabrina Merlo

SALES & ADVERTISING

SENIOR SALES MANAGER
Katie Dougherty Kunde
katie@oreilly.com

SALES MANAGER
Cecily Benzon
cbenzon@oreilly.com

SALES MANAGER
Brigitte Kunde
brigitte@oreilly.com

CLIENT SERVICES MANAGER
Sheena Stevens
sheena@oreilly.com

SALES & MARKETING COORDINATOR
Gillian BenAry

MARKETING

SENIOR DIRECTOR OF MARKETING
Vickie Welch
vwelch@oreilly.com

MARKETING COORDINATOR
Meg Mason

PUBLISHING & PRODUCT DEVELOPMENT

DIRECTOR, CONTENT SERVICES
Melissa Morgan
melissa@oreilly.com

DIRECTOR, RETAIL MARKETING & OPERATIONS
Heather Harmon Cochran
heatherh@oreilly.com

BUSINESS MANAGER
Rob DeMartin
rdemartin@oreilly.com

OPERATIONS MANAGER
Rob Bullington

MAKER SHED EVANGELIST
Michael Castor

COMMUNITY MANAGER
John Baichtal

ADMINISTRATIVE ASSISTANT
Suzanne Huston

PUBLISHED BY
O'REILLY MEDIA, INC.
Tim O'Reilly, CEO
Laura Baldwin, President

Copyright © 2012
O'Reilly Media, Inc.
All rights reserved.
Reproduction without
permission is prohibited.
Printed in the USA by
Schumann Printers, Inc.

Visit us online:
makezine.com

Comments may be sent to:
editor@makezine.com

CUSTOMER SERVICE
cs@readerservices.
makezine.com

Manage your account online,
including change of address:
makezine.com/account
866-289-8847 toll-free
in U.S. and Canada
818-487-2037,
5 a.m.–5 p.m., PST

Follow us on Twitter:
@make
@craft
@makerfaire
@makershed
On Facebook: makemagazine

CONTRIBUTING EDITORS
William Gurstelle, Mister Jalopy, Brian Jepson, Charles Platt

CONTRIBUTING WRITERS
William Abernathy, Thomas J. Arey, Peter Bebergal,
Wendy Becktold, Sean Bonner, Chris Connors,
Steve Crawford, Gus Dassios, Stuart Deutsch, Jeremy Elson,
Gregory Gage, Reed Ghazala, Saul Griffith, Matthew Gryczan,
Jon Kalish, Mark Karpel, Laura Kiniry, Robert Knetzger,
Kris Kortright, Ben Krasnow, Andrew Lewis, Frits Lyneborg,
Timothy Marzullo, Michael Mauser, Forrest M. Mims III,
Lina Nilsson, Meara O'Reilly, Charles Platt, Rick Schertle,
Dean Segovis, Ariel Levi Simons, James Peyer,
AnnMarie Polsenberg Thomas, Joe Sandor, Ariel Waldman,
Robert H. Walker

CONTRIBUTING ARTISTS & PHOTOGRAPHERS
Roy Doty, Nick Dragotta, Timmy Kucynda, Juan Leguizamon,
Tim Lillis, Rob Nance, Kathryn Rathke, Damien Scogin

ONLINE CONTRIBUTORS
John Baichtal, Michael Castor, Michael Colombo,
Chris Connors, Collin Cunningham, Adam Flaherty,
Nick Normal, John Edgar Park,
Sean Michael Ragan, Matt Richardson

TECHNICAL ADVISORY BOARD
Kipp Bradford, Evil Mad Scientist Laboratories,
Limor Fried, Joe Grand, Saul Griffith, William Gurstelle,
Bunnie Huang, Tom Igoe, Mister Jalopy, Steve Lodefink,
Erica Sadun, Marc de Vinck

INTERNS
Eric Chu (engr.), Craig Couden (edit.),
Max Eliaser (engr.), Gunther Kirsch (photo),
Ben Lancaster (engr.), Miranda Mager (sales);
Brian Melani (engr.), Tyler Moskowite (web),
Paul Mundell (engr.), Nick Raymond (engr.),
Daniel Spangler (engr.)

CONTRIBUTORS

Sean Bonner (*Drive-By Science*) has "never been a fan of asking for permission to do anything, so pretty much every project has resulted from just doing it. To mixed results." He lives in Los Angeles but spends "a heck of a lot of time in Tokyo," has a 2-year-old son named Ripley, is happily married to Tara Brown, has a cat and a dog, and loves almond milk lattes and wings from Veggie Grill. He's currently working on a bunch of photo projects, a book about getting rid of personal possessions, a music project exploring the lost thrill of discovery, perfecting a banana-coconut-kale smoothie recipe, and trying to learn how to use a straight razor without killing himself.

Ben Krasnow (*My Scanning Electron Microscope*) works on Top Secret projects at Valve Corporation. For his previous day job, he built computer peripherals for MRI machines. His company, Mag Design and Engineering, sold these devices directly to researchers at academic institutions who use them to publish scientific papers in peer-reviewed journals. After work, he spends time on many different types of projects that usually involve circuit design, machining, material selection, and general fabrication/hacking. His favorite place to be is his home workshop. He's also writing a book that documents his favorite projects.

Rick Schertle (*Rocket Glider*) has taught middle school language arts and history for the past 19 years and is trying to practice what he teaches by writing for MAKE. Along with his wife and young son and daughter, he road-tripped across Europe and North Africa, couch surfing along the way. Following his interest in all things that fly, his next project is building the flying Towel from the last issue of MAKE. Time rich but cash poor, Rick dreams about his next ultra-budget overseas sabbatical with his family.

Dean Segovis (*Fetch-O-Matic*) is a self-taught hardware hacker/problem solver/inventor. He built his first crystal radio in 1973 at the age of 13. It didn't work because he used a real cat's whisker! (Thanks Kitty!) He loves to relax by taking whatever is at hand and seeing how he can repurpose it. Dean is a tinkering/inventing type at heart and is currently working on a proprietary leveling device that he's hoping to take to market. He makes a living as a European auto technician, where he gets to troubleshoot and problem-solve every day. He lives in North Carolina in a nice, small Southern town with his two dogs, three cats, and three chickens!

Lina Nilsson (*DIY Lab Equipment*) is a biomedical engineer at the University of California, Berkeley. Previously, she has worked at a fishery in Norway and on a vineyard in Germany. She can thus not only purify proteins, but also both clean salmon and remove weeds at alarming speeds. Lina believes in the power of open spaces: she ski mountaineers the Sierras and volunteers as a backpacking guide in Alaska. She also firmly believes in the power of open science: academic science can and should be fundamentally transformed by adopting citizen science approaches of openness, crowd sourcing, and flexibility.

Nick Dragotta (*Howtoons*) is proud to have been in MAKE since Volume 01! He lives in Alameda, Calif., with his wife and fellow Howtoons designer, Ingrid, their son Leo, and a cat named Tiger. While being a comic book artist has its frustrations (the first *Howtoons* book was shelved in Adult Science at a major bookstore chain because it was "too dangerous"), he's really excited about the projects he's working on: a *Howtoons* energy literacy book, *Howtoons* for mobile devices, and the first *Howtoons* summer camp in Madison, Wis. What does he do for fun? "I get to make comics for a living, but the family bike rides might just take the cake," he says.

You design an electronics project.
We'll turn your bright ideas into cash!

HAVE A GREAT PROJECT IDEA?

Design an electronics project at ClubJameco.com, identify the components, write step-by-step instructions, and that's it. We'll do everything else!

Start Earning Now!
www.ClubJameco.com

Three Test Tubes and the Truth

CYBERPUNK. STEAMPUNK. DIESELPUNK. The Atari Punk Console. "What's with all the punk?" people frequently ask. What on Earth could be the connection between rebellious, angst-ridden teen music and various geeky subcultures and styles of making?

Mainly, it's about going for it. Doing first, questioning yourself later. Stories abound of folks like a young Elvis Costello seeing the Sex Pistols on TV and thinking: *Hey, they kind of stink. I can do better.* And with that, many a new-generation musical superstar was born. Punk literally gave them "permission to play."

If I didn't love Massimo Banzi already, I did after opening his book, *Getting Started with Arduino*, and finding an illustration from an old punk rock zine with a crude drawing of the guitar chords A, E, and G, and the instruction: "Now form a band." As used to be said of punk rockers (and country musicians before them): they had nothing more than "three chords and the truth."

So this brings us to our issue's theme, Punk Science. All around the world, just as individuals and small groups are discovering the joys of tinkering with the made world, citizen scientists are also getting together to understand more about the natural world and the realms of science.

You can't bring up the subjects of citizen science, biohacking, and exploring genomics without cautious sorts worrying about engineering environmental disasters or Homeland Security showing up to question the presence of your lab equipment. While we at MAKE value safety and being responsible, and all the citizen scientists we know feel the same way, in the face of over-reactive caution, we shout: Punk rock! Permission to play!

Be smart, do your homework, work safely, but get out there and do it! Peer under rocks, into microscopes, and up into the heavens. Extract your own DNA. Record and analyze the processes of your body. Play with your food. Experience science in your own hands, not just as cable-borne entertainment!

> Peer under rocks, into microscopes, and up into the heavens. Extract your own DNA. Play with your food.

In this issue's special Punk Science section (starting on page 36), we look at making your own lab equipment, offer tips for effective citizen science, show you how to throw off high-voltage sparks and do kitchen-table biotech, and even tell you how one maker built his own scanning electron microscope for under $2,000. Then there's the RoboRoach [shudder]. And plenty of other fun projects to keep you in your lab coat.

"Punk" also implies simplicity, a big bang for the buck. In our cover project, Dean Segovis details how to use a motorized, spring-loaded "whacker arm" to create an automatic dog ball launcher. The dog plays fetch with himself! Further keeping things low-fi, Rick Schertle brings the classic Rocket Glider back to life in a project that's as fun to build as it is to launch high into summer skies.

And if all these punk rock analogies have you in the mood to kick out the jams, check out Bill Gurstelle's Sound-O-Light speakers, made from clear PVC pipe. They deliver a cool look, flashing lights, and surprisingly good sound.

Whatever you decide to build and explore, don't wait for permission. The world is yours to jam. Go ahead. Do it. Be a punk! ◪

Gareth Branwyn is editorial director at MAKE.

Automation UPStart, Bacteria Battery, and Stone Grinding

📷 The chart in your article "DIY Home Automation" [Volume 30, page 42] says that "Setting up Universal Powerline Bus requires expensive software tools, usually purchased by a contractor." That's incorrect. Setting up UPB requires software called UPStart, which is available free to anyone and is all you need to set up many installations. UPStart even has signal strength and noise meters built in, something that most protocols need expensive hardware devices to do.

Of course if you want a really fancy setup, you'll want to purchase a home automation software package, but this is true for all the other protocols as well. In fact, all the fancy software packages I've investigated defer significant chunks of their capabilities to UPStart. I'm just a UPB user, not associated with any UPB manufacturer.

—*Steve Kocan, Frederick, Md.*

✉ In Volume 30's "Toys, Tricks, and Teasers" article [page 152, "Things You Can't Make"], Mr. Simanek asks: "Can you cut a triangle from a flat piece of paper that has exactly equal sides but unequal angles?" As a matter of fact, it's easy to do.

I think what he had in mind as an impossible task is to cut a triangle that has exactly equal sides but unequal angles from a flat piece of paper. It's the geometry of the triangle, not of the piece of paper, that is constrained.

Mathematics requires unambiguous expression of ideas, and the formal mathematical language does it very well. Most of the time, with a little care, it is also possible to avoid ambiguities in English.

If you guessed that I'm a stickler about math, you're right. My passion is K–12 math education and I'm working on making the

learning of math more tangible. Physical objects that appear to be impossible due to their 2D perception are a great tool for teaching several mathematical topics. Until I saw the photos of Lego constructions in your article, I thought you had to be a skilled woodworker to accomplish this. This example is wonderful. Keep up the good work.

—*Uri Geva, San Mateo, Calif.*

PROJECTS EDITOR KEITH HAMMOND REPLIES: Uri, you're right, those two clauses should be swapped for clarity. Thanks for the sharp eyes!

✉ Regarding the "Bacteria Battery" article [Volume 30, page 42]: I built my Mudwatt months ago. I was pleased that the article talks about follow-on projects but disappointed that it doesn't provide sources for supplies

L–Dopa (top)

or address design issues in more detail. The parts list essentially is MAKE saying, "Buy the kit from us." Doesn't that make this a really long advertisement? MAKE should be facilitating hacking kits or building new designs, not just buying them.

—*John Grout, Rome, Ga.*

EDITOR-IN-CHIEF MARK FRAUENFELDER REPLIES: John, thanks for your note. It's a tradition for how-to magazines to offer kits, and we consider it a service to the reader. Quality kits come with the correct components, they're complete, and they're usually less expensive than buying the parts separately. They give the builder a good shot at a successful build. Kits aren't for everybody, but for those who don't have the time or money to start from scratch they're terrific.

The scratch-build instructions in this article seem adequate, but we know there's always room for improvement. We encourage you (and other readers) to improve upon it in our Make: Projects wiki at makeprojects.com/project/b/2106.

✉ When I was a kid (back in the 1970s) I sent away my 99 cents for a 6-foot Frankenstein's Monster. I received my monster but not the beast I was expecting. Instead, it was an image of the monster printed on a sheet of plastic, sort of a large plastic trash bag. I certainly learned my lesson.

However, I disagree with your comments about the 9-Foot Hot Air Balloon kit [MAKE Ultimate Kit Guide, "Comic Book Kits That Suck"]. I ordered one, as did my entire tribe of Indian Guides (a father-son organization sponsored by the YMCA). Each father and son built their balloons at home, and we launched them during one of our camping excursions.

The balloons worked fantastic! It was fun to see which of our balloons would be caught by the wind and carried for miles and which would get eaten up by the trees. It's a shame they were one-time use only, but at $2, it was well worth the experience. I wouldn't mind making a few more with my own kids.

—*Steve Frey, Dearborn Heights, Mich.*

PROJECTS EDITOR KEITH HAMMOND REPLIES: Steve, we think you and your kids will enjoy our "$4 Hot Air Balloon" project in MAKE Volume 30.

✉ I just wanted to let you know how much I enjoy the magazine. I recently finished building the Bubblebot [Volume 28, "Gigantic Bubble Generator"]. I could not have done it without the kindness and patience of the author, Zvika Markfeld. The articles are great, but combine that with a truly kind, nurturing tutor and mentor like Zvika and you can't miss.

—*Lawrence Freundlich, New York, N.Y.*

✉ I've been making stone bowls for 15 years, and there's more to it than what's shown in "Carve a Stone Bowl" [Volume 24, page 168].

For one thing, use a turbo cup wheel. Always wear body, eye, and hearing protection; I had a cup wheel come apart, and at 10,000rpm it will tear right through clothing. I use 16", 14", 12", 10", 7", and 4" blades. Don't buy cheap blades; that two-for-a-nickel stuff ain't worth the time it takes to mount it.

I spent a lot of time and money doing it wrong, ruined several thousand dollars' worth of equipment, and bought lots of crappy stuff that didn't do the job. There are good blades and polishing equipment at reasonable prices.

—*Tim Boehm, Tillamook, Ore.*

DON'T DO THIS

MAKE AMENDS

In Volume 30's "Yakitori Grill" project (makeprojects.com/project/y/2063), we purchased the deep cake pans at webstaurantstore.com, but we printed an incorrect URL.

In Volume 30's "Indestructible LED Lantern" project (makeprojects.com/project/e/2092), author Steve Hoefer is pictured wearing gloves while using a drill press, which is a safety hazard. "Gloves are a no-no around rotating tools," Steve admits. Thanks to reader Tim Kemp for pointing out the error.

In Volume 30's Howtoons project [page 156, "The Measure of a Man"] the 7" measurement is incorrect. It should be measured from the left outside edge to the right edge of the smallest rectangle. Thanks to reader Thomas Zink of Springfield, Ohio, for spotting the error.

MAKER'S CALENDAR
Compiled by William Gurstelle

Our favorite events from around the world.

World Maker Faire
Sept. 29–30, Queens, N.Y.
Each year, the World Maker Faire gets bigger and better, and this year will be no exception. See incredible projects and interact with artists, crafters, engineers, scientists, and everyone else in the East Coast maker community.
makerfaire.com

AUGUST

» Pyrotechnics Guild International 2012 Convention
Aug. 11–17, La Porte, Ind.
Some of the world's best and most technically adept pyrotechnicians will gather just south of Lake Michigan to celebrate all things that go whoosh and boom.
pgi.org/convention

» The Great Dorset Steam Fair
Aug. 29–Sept. 2, Dorset, England
With an incredible 200,000 visitors, The Great Dorset Steam Fair is quite likely the largest steam engine show on the planet. The show includes displays and demonstrations of steam engines, vintage cars, crafts, and a music program. gdsf.co.uk

» Gogbot
Sept. 6–9, Enschede, Netherlands
The idea behind the Gogbot festival is to take high quality multimedia, art, music, and technology out of museums and make it accessible for a broad audience by presenting it in a public space. The event also includes lectures, a film program, and concerts.
gogbot.nl

SEPTEMBER

» Next Big Idea
Sept. 14–15, Los Alamos, N.M.
Los Alamos has seen its share of big ideas. Building on that tradition, the city is holding a two-day festival that provides a venue for scientists, technologists, innovative artists, and inventors to share their ideas with attendees and each other. nextbigideala.com

» Mother Earth News Fair
Sept. 21–23, Seven Springs, Pa.
Fairgoers explore the possibility of a greener, more sustainable world through demos, lectures, workshops, and exhibits. More than 200 presenters and vendors will survey everything from sustainability and gardening to baking and the DIY lifestyle.
motherearthnews.com/fair/sevensprings.aspx

OCTOBER

» Outer Banks Stunt Kite Competition
Oct. 13–14, Kill Devil Hills, N.C.
It's literally cutting edge entertainment at this festival, where novices and pros make and fly some of the best and most unusual kites found anywhere.
makezine.com/go/outerbanks

» Colorado Springs Cool Science Festival
Oct. 13–20, Colorado Springs, Colo.
Local schools, businesses, entrepreneurs, scientists, and engineers collaborate to offer a potpourri of science-related activities and programs over the course of a week. csscp.org/csfest

✳ IMPORTANT: Times, dates, locations, and events are subject to change. Verify all information before making plans to attend.

MORE MAKER EVENTS:
Visit makezine.com/events to find classes, fairs, exhibitions, and more. Log in to add your events, or email them to events@makezine.com. Attended a great event? Talk about it at forums.makezine.com.

Andrew Kelly

Not Failure

WHEN I MOVED TO ST. PAUL THERE WERE two things that I was really excited about: my new job as an engineering professor, and the circus school a short distance from my house. Within months, I was hooked and taking classes in flying trapeze, Spanish web, and whatever other equipment I could get access to. Thus, I was spending a lot of time around spinning, bouncing, and turning people.

I remember thinking how it was like being in the world's coolest physics lab. Suddenly I was the mass that was swinging from a pendulum (flying trapeze), bouncing on a spring (bungee trapeze), or in a transforming coordinate frame (German wheel, which is like a human-sized hamster wheel that isn't locked in place). This seemed like the perfect time to combine work and play, and a few years later I found myself teaching a dynamics course where the experimental labs took place at the circus school. Think "daring young engineers on the flying trapeze."

Fun as the class was, one of the most amazing parts for me came on the final day. Rather than give my students a written final, I asked them to perform a circus for middle school students, in which the science behind the acts was explained. They took the challenge, and performed a great show.

One act consisted of a student swinging on a low-casting rig (a small version of a flying trapeze). As he swung, his partners explained conservation of energy, discussing how potential energy is converted to kinetic energy. As the amplitude of his swing decreased, he demonstrated how beating (kicking at specific points in the swing) allowed him to increase the swing amplitude. But where did his body get the energy to perform these kicks? Chemical energy from food that he'd consumed.

Another group of students discussed inertia, center of mass, and gravity, showing how they could control a German wheel by changing where they stood in, or on, the device. In the middle of this demonstration, in front of a room full of middle school students, the wheel fell over with one of my students inside of it. Rather than get flustered, this young woman stood up, looked at the students, and said that in engineering things don't always work out the way you want them to, but you get up, brush yourself off, and try again.

Think "daring young engineers on the flying trapeze."

With all eyes on her, this is exactly what she did, and she performed a perfect series of "cartwheels" in her wheel. Honestly, of all the lessons that the young audience heard that day, I suspect the most important one was this impromptu demonstration of the power of being willing to try again.

I'm sure every maker can relate to this moment and to this lesson. You think the project is going perfectly and then the part breaks, the light fades, or something starts spewing smoke.

People extol the virtues of failure, and how much you can learn from it. I agree, but prefer to focus more on resilience. It's not failure itself that leads to success, it's the willingness to pick yourself (and your project) back up in hopes of getting it to work. If that's what you do, those early hiccups weren't failures, they were rough drafts. To me, it's only truly a failure if you give up. ◪

AnnMarie Thomas, mother of two young makers, is the executive director of the Maker Education Initiative (makered.org).

Made On Earth
Reports from the world of backyard technology

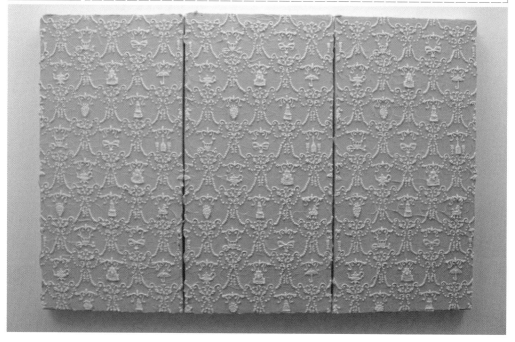

Queen of Caulk

Seated on the Aar River, Solothurn calls itself Switzerland's most beautiful baroque city. So naturally, when multitalented California artist **Tramaine de Senna** was offered a residency there, she was inspired to create works that are a twisted tribute to this elaborate style.

In her studio overlooking the Aar, de Senna buttoned up her lab coat, donned her latex gloves and face mask, and spent the next three full months clocking about 800 hours working on her "I Love Caulk Frosting" series. Duality and reappropriation of mundane materials are central themes in much of de Senna's work, and industrial caulk fits both bills, visually resembling tasty sugar frosting but most often toxic in content. The 16 total works range in size from a triptych titled *Santa Claus, Schmutzli and Satan* totaling roughly 9' tall by 6' wide (which took 216 ounces of caulk), to nine small squares, like *Snow Queen*, measuring about 7" by 7".

De Senna's process was straightforward but laborious. She began by drafting traditional baroque patterns on transparent film, then infusing them with benign images ranging from pretzels to pine trees, randomly interspersed with unexpected dark counterparts like skulls, mudflap girls, and exploding-head squirrels. Starting with canvases coated with a layer of gesso topped with a layer of caulk, she projected the patterns onto the canvases, then painstakingly caulked the designs in place with a manual caulking gun.

She used silicone for most pieces, preferring its flexiblity after hardening and easy cleanup. She used painted acrylic caulk for *Laudrée Royale,* since green caulk is hard to find, and for *Swiss Milk Chocolate Delight* because acrylic brown looked more "chocolatey." Who knew caulk came in so many flavors?

—*Goli Mohammadi*

➕ tramainedesenna.com

Tramaine de Senna

Motor Mashup Master

Randy Grubb has one sweet job: working from his Grants Pass, Ore., garage, he builds hot rods. His rides, however, are far from ordinary. No small-block Chevy engines or deuce coupe chassis litter his yard. He builds from supersized truck, jet, and tank parts.

The son and grandson of dentists, Grubb seemed destined to continue the family trade until a glassblowing demo in college led him astray. Using French "lamp work" techniques, he built brightly colored glass sculpture paperweights, which earned him a good living for 20 years. For fun, he built cars.

At age 40, Grubb decided to take a year off to work on a giant car. The next month, 9/11 occurred, cratering the high-end collectible market. "There was no career left to go back to," he says. He poured his life savings and a year of work into the *Blastolene Special*, a 9,500-pound roadster built around a 29-liter, 910hp M47 Patton tank engine. Flush with

new owner Jay Leno's cash, he found a new career: extreme cars, built on spec.

With the success and acclaim that the *Blastolene Special* brought him, Grubb continues to work long hours on new rolling sculptures, winning awards and (he hopes) a place in the pantheon of great coachbuilders. His latest creation, the *Decoliner*, mashes up a 1973 GMC motorhome and a 1955 White Motor Company cabover truck with Buck Rogers styling cues. With its polished aluminum finish, nautical portholes, and rooftop fly bridge for land-yachting, the *Decoliner* is a truly unique ride.

"I consider myself really lucky," Grubb says. "I can spend 3,000 to 6,000 hours on a single project in a culture where you can hardly get a three-second sound bite in. That's what it takes to make something special."

— *William Abernathy*

➕ blastolene.com

Randy Grubb

Achieving Balance

Tired of walking to the supermarket from his dorm, 21-year-old MIT computer science and electrical engineering student **Stephan Boyer** decided he needed a quicker way to travel.

"I wanted a personal transporter that could get me around campus but be as small and light and fast as possible," he says. Five months later, Boyer had Bullet, a self-balancing electric unicycle that reaches a top speed of 15 miles per hour and is the envy of his peers.

Boyer built Bullet with $1,000 and endless ingenuity. After teaching himself welding and some mechanical engineering, he got to work collecting materials: a basic fork to hold both a moped wheel and motor, welded-on pedals to provide foot support, and a top seat. "I didn't really plan out the [entire design], so the battery and electronics are held on with zip ties," he says. Bullet also features a kill switch for swift deactivation. He says the design could be built for $600.

For balance, Bullet employs an onboard computer and two sensors: a gyro and an accelerometer. The former measures Bullet's rotation speed and the latter determines its acceleration due to gravity. Using this data, the computer can detect the angle of the unicycle and prevent it from leaning too far forward or backward.

"Because Bullet only has one wheel, you still have to balance side to side," says Boyer, who does this by twisting his hips to turn and occasionally flailing his arms for stabilization, in the same way a flying squirrel uses its tail. Riding it several miles daily, Boyer became somewhat of a pro at handling the vehicle, but he stresses that it takes both skill and patience. "My number one rule when riding Bullet is that things in the road are *always* bigger than they appear."

— *Laura Kiniry*

stephanboyer.com

Slow Art Pottery

Nestled in the cozy woods of Minnesota lies a small yet efficient pottery studio on the grounds of Saint John's University. The Saint John's Pottery Studio, run by master potter **Richard Bresnahan**, is nationally respected. The studio is like the art world's version of the Slow Food movement, where conscious attention is given to where and how materials are used, in a sustainable manner.

Bresnahan, his apprentices, and the studio are part of a local system that embodies Benedictine values. In 1979, he returned to his alma mater from an apprenticeship in Japan and was invited to build a pottery program using techniques he learned. All of the studio's materials are locally sourced, many from the Saint John's Abbey grounds. Early on, he nabbed 18,000 tons of clay from a nearby construction site, which he still uses today. He also has access to kaolin clay from decomposed granite in a nearby quarry. The clay is processed on site by apprentices using salvaged factory machines and gray water.

In the fall, after a year's worth of preparation, they hold a kiln-lighting ceremony with the monastery and community to fire the Johanna kiln, the largest wood-fired kiln of its kind in North America (it holds 12,000 pieces), for ten days straight, followed by ten to 14 days to cool and unload. It requires a team of 40 to 60 people to successfully fire, with temperatures reaching 2,500°F. Deadfall wood collected within Saint John's Arboretum serves as fuel, which adds a unique element of beauty, the ash forming a subtle glaze.

"What are the moral and ecological issues facing our world today, and how are artists uniquely equipped to tackle these issues?" asks Bresnahan. Saint John's Pottery is a holistic response to this inquiry.

—*Joe Sandor*

➕ csbsju.edu/pottery

Joe Sandor

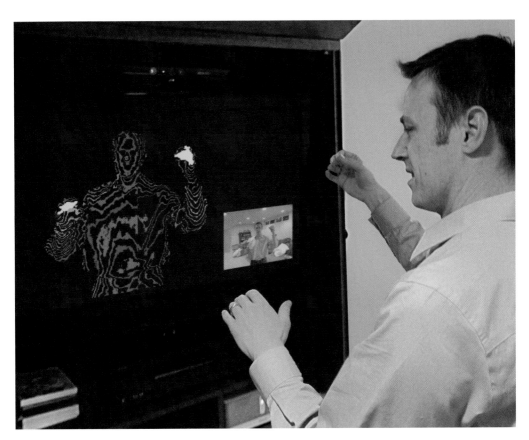

Maestro of the Wii

The theremin is a relatively modern musical instrument. Invented in 1920 by Russian physicist Léon Theremin, it's unique in that it's played without being touched. Pitch and volume are determined by the proximity of the hands to two antennae. The resulting sounds are eerie and reminiscent of mid 20th-century sci-fi films like *The Day The Earth Stood Still*.

Several years ago, **Ken Moore**, a Seattle-based user experience designer for Google, built a theremin that adheres to the traditional touchless play, but instead of using radio frequency oscillators to sense movement, he used a Nintendo Wii Controller. The Wii detects infrared light emanating from the LEDs embedded in the index fingers of a pair of leather gloves worn by the musician. The position of the player's hands controls the sounds: left hand for volume, right hand for pitch. The information gets transmitted via MIDI to a synthesizer, which creates the actual sound.

Two years later, Moore, 43, built another theremin — this time using a Microsoft Kinect to detect hand motions, mapping them to correspond to pitch (right hand, z-axis), volume (left hand, y-axis) and modulation (left hand, x-axis). Onscreen, the mapped body displays as a pulsing rainbow, its brightness and colors fluctuating based on volume and pitch.

After building both, Moore concluded that the Wii version is superior because the Kinect controller has significant latency and the Wii handles slides between notes better, due to a higher data-sampling rate.

And because both theremins use a synthesizer to create sound, "I'm not restricted to the sine wave sound of a traditional theremin, so the sonic possibilities are endless," muses Moore. "My father-in-law, a dentist, suggested I should make it sound like a dental drill."

—*Laura Cochrane*

➕ kenmooredesign.blogspot.com

Kali Sakai

Made in Sector Zero-Zero-One

Free-associate the phrase "starship *Enterprise* coffee table," and generally, words like "classy" and "elegant" do not come first to mind. But woodworker **Barry Shields'** actual starship *Enterprise* coffee table manages to be both those things, without sacrificing the "wicked" and "cool" factors inherent to the concept. Come 2375, it will probably grace the commandant's office at Starfleet Academy; for now, it belongs to one very lucky professor at the University of Connecticut.

Shields, father of two, lives in Sevierville, Tenn., and works as a technician by day. His *Enterprise* table was built over a month of nights and weekends in his dad's furniture restoration shop and sold shortly after listing on Etsy earlier this year. It has since attracted wide attention online, including praise from *Star Trek* luminaries Rod Roddenberry and George Takei, and Shields is hard at work on new orders.

The ship is modeled in fine hardwoods — ash, cherry, and poplar — and the top is a shield-shaped piece of glass 3 feet wide by 4½ feet long. Shields says cutting the glass was among the most difficult parts of the project, involving expert assistance, a handheld cutting wheel, and "a lot of prayer."

Astute Trekkies will recognize the ship as the ill-fated NCC-1701-C, which Shields chose mostly for practical reasons: the bridge and nacelles of *Enterprise* C are closer to the same height than those of any other ship to bear the name. Even so, to level the tabletop he had to use small risers on the nacelles and mount the bridge dome above the glass. The gentle arc of the base is intended to suggest a planet's horizon below, but many have pointed out its resemblance to a Klingon bat'leth.

And we know, gentle reader, that we don't have to tell you what that is.

— *Sean Michael Ragan*

Barry Shields

Real Cars, 8-Bit Frogs

In honor of *Frogger's* 30th birthday, New York-based advertising creative director **Tyler DeAngelo** put a new spin on the classic video game. He calls it *5th Avenue Frogger*, and it uses data on actual street traffic that the player's frog must avoid in order to cross the street. To accomplish this, an internet-connected webcam transmits a video feed from over Fifth Avenue in New York City. The machine connects to the video feed, tracks the vehicles' positions, and translates them into 8-bit vehicles driving along a virtual street. The game is housed within an arcade cabinet from the original version of the game, and it can be switched back into classic mode in case there's too much traffic on the street.

DeAngelo worked with **Ranjit Bhatnagar** and **Renee Lee**, who helped bring his idea to fruition. Bhatnagar retrofitted the arcade cabinet with a PC, wrote the code for the new version of the game in C++ using the Simple DirectMedia Layer library, and used the OpenCV library to analyze the traffic video. Seamlessly integrating old and new technology proved to be the most challenging part of the build. Simply getting the old monitor to display the PC's video signal required a custom circuit board.

While the game cabinet usually sits in DeAngelo's office, he occasionally wheels it out onto the streets of Manhattan to let New Yorkers play. "I dramatically underestimated the challenges of maneuvering and powering an arcade machine on the street," he says. "However, once we were able to get it running, the response was really positive."

He doesn't take all the credit, though: "I think what I did is similar to remixing a hit song. ... I gave it a little tweak so that it's fun to experience again in a new way."

—*Matt Richardson*

⊞ 5thavefrogger.com

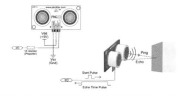
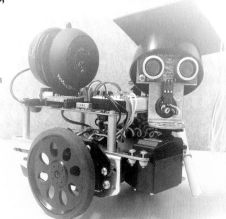
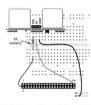

The Year of Cheap Robots That Make Cheap Robots

I'VE BEEN THINKING A LOT ABOUT CNC machines lately. My workshop runs on them and they are hugely productive: insert CAD file, out comes usable parts. My favorite for its sheer speed and versatility is our laser cutter; it's fast, can make big things, and can use real, strong materials.

I'm working with Jonathan Ward and Mike Estee at Otherlab to build a CNC machine with the capacity of a laser cutter but the cost of a desktop inkjet. We're building it under the DARPA MENTOR program; the goal is very low-cost, open source machines that can be placed in classrooms. Students would gain access to CNC machines that can produce a variety of useful machines, mechanisms, and hands-on lessons.

I believe DARPA, like me, is concerned about the future of manufacturing and where the manufacturing base is. The last few decades have seen manufacturing magnetically move to geographic places with low wages — so much so that it's now difficult to get high-quality things made in the United States. I don't think this will last; I believe in a future with technologies like Nike's Flyknit, a highly automated, almost-CNC way of making shoes using only a few parts and a few people.

If there's a future for American manufacturing, it's in this type of automation and robotics. But if manufacturing jobs are going to be replaced by CNC, what are we all going to work on? We should be building a generation of machine designers, product designers, and market designers to tell those machines to build the machines we need to make the things we need for the markets that need them. A country full of end-market consumers isn't sustainable.

It's exciting to see different communities, large and small, pushing manufacturing toward this future of highly specific CNC machines. Every day new projects pop up around building open source CNC machines, 3D printers, laser cutters, mills, and lathes. This year we're making a machine for high school students, but we hope many other machines are produced to complete the low-cost CNC workshop of the future.

My real dream is that we improve these

A country full of end-market consumers isn't sustainable. We need machine makers.

machines with every generation, and that each community leverages the progress in all the others. The knowledge of how machine tools are built and more importantly, designed, needs to be much, much more accessible. Makers of new CNC machines should specify the tolerances, repeatability, and capacities of their machines. We need to measure which ones are hitting their specs, and improve each other's machines with rigorous machine design so that our specs keep improving.

Let's share components, drivers, software, firmware, and hardware and make it a cheaper sandbox for all to play in. This is how the open source CNC communities can make a serious dent in changing the way things are made and the way we educate the people who will make the machines that make things in the future.

I hope the low-cost CNC we produce for the DARPA program helps spur this process, giving students firsthand experience with machines that make machines, and enabling them to design the plethora of CNC machines of America's manufacturing future. ◪

Saul Griffith is chief troublemaker at otherlab.com.

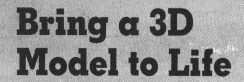

Meet the Arduino Leonardo

IT'S NO SECRET THAT MAKERS OF ALL skill levels are adopting Arduino — the open source ecosystem of microcontrollers, add-on "shields," sensors, and code — to bring all kinds of DIY projects to life. In this column I'll bring you cool new stars in the Arduinoverse, and I hope you'll tell me about great stuff you're finding at whatscool@makezine.com.

Arduino Leonardo

The Arduino Leonardo (makershed.com/leonardo) made its first retail appearance at Maker Faire Bay Area 2012. If you were lucky enough to attend, you may have picked up one from the Maker Shed store. If not, don't worry — now you can get them online.

Leonardo is a very different kind of Arduino, yet it still works as elegantly as previous versions of this amazing little microcontroller. At its heart is the ATmega32U4 chip, which eliminates the need for a secondary processor for USB communications. This not only lowers the price of the board, but also allows it to be used as a virtual keyboard or mouse! Imagine the new possibilities for letting your projects interact with computers.

Mintronics: Menta

This kit was designed in New York City by open source hardware pioneer Adafruit Industries, in partnership with MAKE and Maker Shed (makershed.com/menta). It's an Arduino-compatible microcontroller that fits perfectly inside everybody's favorite enclosure, the mint tin — complete with onboard prototyping area and shield compatibility.

And check out the 3D-printed Menta enclosure created by Tod Kurt (aka todbot). Tod uploaded his design files to Thingiverse (thingiverse.com/thing:23809) so you can easily download and modify the design. It's perfect if you want a little more height for stacking shields or adding jumper wires, sensors, or battery packs to your Menta.

Arduino Wireless Proto Shield

This new shield (makershed.com/mksp13) allows your Arduinos to easily communicate with each other and with your computer from 100–300 feet. It was designed to use Digi's XBee radio modules, but any wireless module with the same footprint will work. Arduino also sells a version with an SD memory card slot.

Arduino Ethernet

This microcontroller integrates an Arduino and the popular Arduino Ethernet Shield. By downloading and using the Ethernet library in your Arduino sketches, you can connect all kinds of projects to the web, like Matt Richardson's Snail Mail Push Alert system for his iPhone (makezine.com/go/snailpush) and Chris D'Angelo's tkts Ticker Tape gadget that researches and prints out Broadway theater ticket discounts (makezine.com/go/tixticker).

Now you don't need a separate shield for web connectivity — just plug and play with Arduino Ethernet (makershed.com/mksp9). ▨

Marc de Vinck is a father, professor, and maker and breaker of things.

Maker

TINKERING WITH THE MUSIC OF THE SPHERES

Dan Krause's home science museum.

By Mark Karpel

If you're looking for Dan Krause, take the first driveway after the sun. The yellow sphere, ten feet from the road and roughly the size of a large exercise ball, is part of an orrery — a scale model of the solar system — that once stretched for two miles across neighboring fields, backyards, and conservation lands.

Drive in and you'll find an assortment of precise scientific curiosities that spill out of his home, across the surrounding landscape, and into adjacent outbuildings. Krause has filled his property in Amherst, Mass., with handmade devices that explore his lasting love affair with physics, mathematics, astronomy, and time.

Dr. Daniel Krause, Ph.D. — physicist, engineer, astronomical tinkerer, and self-described pagan farmer — worked as an oceanographer and, later, an atomic physicist for the Air Force Office of Scientific Research. He took a position at Amherst College and stayed for 35 years until his recent retirement. Now 67, it's easy to imagine him as an early New England farmer. He rarely uses email, is leery of the internet, and dislikes large cities. However, he spent years scattering subatomic particles off atoms and willingly uses lasers on the property if they'll get the job done.

Krause explores what mathematicians and astronomers once called "the music of the spheres" — a sacred geometry of harmonic proportions discoverable through mathematics. He's fascinated by the connection between the tiny patch of Earth he occupies and the big universe out there. In retirement, he explores what interests him on his own terms. Chuckling, he says, "I don't know what I'm doing, but I'm in a position in life where I can play."

He enjoys answering visitors' questions about how the devices were made and how they work, conversations that often lead to the precision and mystery of the laws that shape our universe. Here are a handful of Krause's unusual creations.

Mark Karpel is a writer and psychologist living in western Massachusetts. He writes about people who think big and get carried away.

Mark Karpel

✎ **COUNTRY SCIENCE** (Top left) Anemometer, which indicates wind speed. (Top right) Rocket made from an old water heater. (Middle) Mirrors bounce the sun's reflections off a barn wall to record solar noon and solar equinox. (Bottom right) Bowling-ball weathervane. (Bottom left) Label for scale model of the sun.

THE SUN
IN THE 1:2,000,000,000 SCALE
MODEL OF THE SOLAR SYSTEM

"I don't know what I'm doing, but I'm in a position in life where I can play."

Orrery

That yellow sphere by the side of the road is exactly one two-billionth the size of the sun. Further back in the yard, three glass globes enclose tiny mounted planets: Mercury, barely bigger than a peppercorn; Venus and Earth, about the size of a raisin. Saturn, since removed, stood a mile away, while Pluto, over two miles away, once graced the yard of a resident who grew so fond of it, he took it with him when he moved. Krause encourages visitors to imagine the sun and planets on his lawn with everything else gone, just the vast emptiness of space, together with the astonishing force of gravity over those staggering distances. It conveys the scale of the solar system better than any small model or book illustration.

Savonius Rotor

Near the orrery, a vertical-axis wind turbine, or Savonius rotor, made from bisected 55-gallon drums, spins when a breeze kicks up. It turns a horizontal rod with a small arm attached underneath. As the arm rotates, it pushes back a large wooden mallet. When released, the mallet swings forward and strikes a rusted metal cylinder, mounted at precisely one-third its length — a "magic number," Krause says — producing an uncanny facsimile of a large, sonorous Japanese bell. The turbine is guarded by a propane-tank pig, one of many whimsical folk art creations Krause sprinkles around the property.

Solar Clock & Calendar

It's easy to miss the chronological device behind Krause's house (see page 31). Two small mirrors, rendered concave by several thousandths-of-an-inch, are mounted high on the wall of an old chicken coop, with three flat ones closer to the ground. Forty feet away, a board, marked off at regular intervals like a ruler, climbs from the ground to the roof of a barn-board wall. The mirrors project reflections of the sun that move across the wall as the day progresses and up and down it as the months go by.

On the solar equinox, reflections from the upper mirrors cross the centerline of the ruler; other lines mark the number of days before and after. The lower mirrors

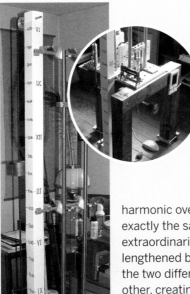

harmonic overtones. When the tubes are exactly the same length, they emit one extraordinarily intense tone. If one tube is lengthened by extending a metal sleeve, the two different tones beat against each other, creating an unearthly sound that, for first-time listeners, is a startling experience. (Note the scorch mark on the wood base, from the sun's reflected rays.)

produce three images of the sun that converge once a day at solar noon, when the sun is at its highest and exactly south of Krause's home. He records and graphs the differences between solar noon and "clock noon."

Thermal Acoustic Generator

Having read that Einstein proposed using sound to generate hot and cold temperatures, Krause decided to use temperature to make sounds.

He mounted two glass test tubes about two inches in diameter on a block of wood and fitted each with a piece of a catalytic converter at one end (seen above, left). Heat can be applied in several ways, but Krause likes using a four-foot bow-shaped mirror (formerly used to cook hot dogs) to focus the sun's rays. When heated on one end, the temperature difference produces oscillations, which we hear as sound.

The tones are absolutely pure, with no

Torsion-Pendulum Clock

In Krause's living room, a clock he made stands about five feet high, with its workings fully exposed. A dumbbell-shaped pendulum rotates horizontally, instead of side-to-side. Other parts include flat torsion springs, an escapement, an assortment of metal pipes and brackets, and a level so sensitive that its tiny air bubble lurches when he slides a piece of paper under one end. Minutes and seconds tick by on small clock faces; a pointer attached to a bicycle chain marks the hour as it descends past numbers on an upright piece of wood.

The clock is extremely temperature-sensitive. A brass block encased in styrofoam contains an internal temperature probe that leads to a digital thermometer, harnessing a modern tool to one of the oldest forms of clockwork known. Krause measures slight variations in the clock's speed related to temperature differences in the room. ◼

BLISSFUL BEDROOMS

DIYers team up to create cool rooms for disabled kids.

By Jon Kalish

For the last three years, a group of New York City DIYers have been doing weekend bedroom makeovers for young people in wheelchairs. Blissful Bedrooms, founded by the husband and wife team of photographer Alex Dvoryadkin and physical therapist Martha Gold-Dvoryadkin, has no paid staff. The volunteer group pays for building supplies and materials used for handmade interior decor with donations solicited online.

Blissful Bedrooms celebrates the passions of the disabled youths while making their bedrooms more functional. Wall murals are painted by a team of artists, and store-bought furniture is sometimes modified to better accommodate the needs of people with limited physical movement.

During a December 2010 makeover at a Bronx public housing project, the Blissful Bedrooms crew built a dressing room vanity into the wall so that Keosha Stukes, who has cerebral palsy, can pull her wheelchair in front of it. An iPad holder was attached to the arm of the wheelchair, for the new iPad Stukes was given as part of the makeover. An engineer improvised a pointer attached to a cap so the young woman can use the iPad's touch screen by moving her head.

Some of the volunteers have serious construction chops. Project manager Adam Seim, 28, worked as a professional carpenter building radio studios before taking a job as a radio engineer at a Manhattan station. Seim was used to a huge woodshop with large power tools and a dust collection system, but now he makes do with a slew of battery-operated

Alex Dvoryadkin

power and hand tools that he carts to the makeover apartments.

"You have to work with the bare minimum of tools," Seim explains as he crouches outside Stukes' apartment, shortening window blinds with a battery-powered jigsaw. "I've learned a lot as a carpenter because of this."

Neighbors on the 10th floor were good-natured about the invasion of the Blissful Bedrooms gang, who used so much of the hallway for construction and fabrication that residents could just barely make it to the elevators and incinerator chute.

For a makeover with a New York Yankees theme, Cory Mahler, who makes her living designing bedding for kids' rooms, sewed a bunch of Yankees T-shirts into a bedspread. That Yankees bedroom is in the home of an 20-year-old named Jesus. (He shares the room with his grandfather, who, the record shall note, happens to be a Mets fan.)

The two headboards in the room are custom-made. One was done in the shape of a baseball and the other as the familiar "NY" in the Yankees logo. They were made from medium-density fiberboard cut on CNC machines at a Brooklyn fabrication shop that builds sets for TV shows and concert tours. Both materials and shop time were donated.

Like the reality TV show makeovers, Blissful Bedrooms has a dramatic reveal on Sunday night, when all the work is completed. The group throws a party attended by lots of young people in wheelchairs.

"They don't get to have these moments, these big milestones," Gold-Dvoryadkin says of the disabled youths. "This is very intense. The fact that so many cared about them and made them the focus of attention for an entire weekend is incredible for them." ◪

➕ For more great before and after pictures, go to blissfulbedrooms.org.

Jon Kalish is a Manhattan-based radio reporter and podcast producer.

ANARCHY IN THE LABORATORY

Real science is happening outside the hallowed halls of high-priced research facilities. Take a look in garages, schools, and hackerspaces and you'll see things like remote controlled cockroaches, crowdsourced radioactivity monitoring programs, lab equipment made from kitchen appliances, biotech research, homebrew scanning electron microscopes, and other full-fledged scientific endeavors.

In the same way that punk rock broke down barriers that previously kept musicians from selling records and playing shows, punk science provides amateur enthusiasts with the skills and tools to make useful contributions to the body of scientific knowledge without having a Ph.D. and millions of dollars in funding. So what are you waiting for — as Johnny Rotten said in 1976, "We don't need permission for anything!"

Juan Leguizamon

3 RULES FOR SUCCESSFUL CITIZEN SCIENCE

Make it measurable. Make it cheap. Make it open.

BY ARIEL LEVI SIMONS

Jorge Luis Borges once wrote, in the very short story "On Exactitude in Science," of a great empire that sought to create a map so accurately detailed that it grew to be as large as the empire itself. Early in my teaching career, and recently out of school, I thought (in a similar, albeit less poetic, fashion than Mr. Borges) about how I was representing science as a high school teacher.

I wrestled with the same anxiety most, if not all, science teachers feel about not covering enough content. I constantly felt rushed to get through enough of the "big ideas," and yet still felt that most of what I was doing existed only in the classroom and vanished the moment my students left for the day.

I felt like I was trying to get my students interested in science from the map I was making, when I should be taking them to the country itself.

Education as Science, Science as Education

My first experience in teaching science by *doing* science happened in late 2009 when I was finishing my first semester teaching an environmental science course at Wildwood School in Los Angeles, Calif. A number of my students told me how relentlessly depressing environmental science felt. Would the entire course be about how our civilization was going to collapse in an ecological catastrophe?

It was a valid question, and after spending that winter holiday thinking how to teach environmental science without inspiring feelings of learned helplessness, I proposed to my class that we find out for ourselves the health of our local environment.

We began an environmental mapping project called TIGER (Technologically Integrated Geotagged Environmental Research). We started measuring and mapping water quality at a number of sites in Los Angeles and soon began measuring air quality and radiation levels along the California coast.

During this time we started connecting with other groups doing citizen science, ranging from studying bird populations to classifying galaxies. We're seeing how citizen science is becoming an increasingly useful tool in research as scientists realize they can get help from a much larger circle than just traditional academia.

From our experience with the TIGER project (scienceland.wikispaces.com/tiger), we've come up with three rules to help classrooms (and anyone else) create, manage, and collaborate on citizen science projects.

Rule #1: Make it measurable.

As any anthropologist with a notebook full of field notes will attest, science doesn't always need to be quantitative to be effective. However, we have stuck with numerical data

Gregory Hayes

✎ **HEALTH CHECK** Wildwood teacher Levi Simons and his students use chemical indicator testing kits to map Los Angeles water quality.

with the TIGER project for two reasons. First, it facilitates comparing data between different monitoring sites and between different dates at the same site. Second, it gives our group a common language to communicate with other schools participating in the TIGER project, and with any outside groups that might want to use our data.

In the field, this has meant using equipment that gives quick numerical results. For example, our water quality data is captured using a variety of chemical tablets that quickly dissolve in test vials, changing color to indicate the concentration of everything from dissolved oxygen to bacteria. At the same time we're testing the water, we also use a pair of electronic sensors: one to measure the concentration of gases such as oxygen and petrochemical vapors, and another to measure weather conditions.

Our goal in taking this data simultaneously has been to look for potential relationships between the atmosphere and water of our local environment. Already we can see periodic ebbs and flows, such as salinity variations caused by the tides.

Students also gain firsthand experience with seeing how educated guesses, errors, and experimental limitations show up in every research project. This answers the complaint, "When am I ever going to use this?"

Rule #2: **Make it cheap.**

Citizen science is frugal science. If you need an outlay of millions of dollars, no one, other than a few large labs, will be able to conduct your research. We have sought to keep our equipment costs low, on the order of hundreds of dollars per school, so as to make our project as accessible as possible. For example, the water quality kits we use cost about $40 for ten full tests. Accessibility is key in creating and managing a project covering a large number of students across a wide geographic area.

With the TIGER project our main costs have been buying the monitoring equipment. Transportation costs are kept low by having each group monitor their local environment and then upload the data to a central website. We also use freely available web-based collaboration software to store and analyze data.

Rule #3: **Make it open.**

Science, whether at a national lab or with a citizen science project, thrives not only on open communication, but on open standards. We treat the procedures of the TIGER project as the software for an open platform.

An open standard means storing and analyzing data using freely available software, but it also means using a set of common and public experimental procedures. We've put our procedures on a wiki, both to ensure uniformity in the type of data collected, and to leave our methodologies open to criticism, which is how science progresses.

Similarly, any citizen science project such as TIGER should also be readily open to expansion. Although we started with just water quality in Los Angeles, we've always used a data framework that can incorporate any environmental metric as long it's recorded with a unique GPS stamp. As a result, we've been able to add more schools and more data types, such as air quality and radiation levels, to our project.

Our openness has also allowed TIGER to connect with other related citizen science projects, such as the community radiation monitoring project, Safecast (*see page 52, "Drive-By Science"*). As Safecast has also been keen on publishing all their data and methods in a public and open format, we at TIGER have had a relatively easy time collecting and analyzing radiation levels for both projects' test sites.

Our collaboration with Safecast has also given TIGER students the opportunity to troubleshoot the Geiger counter hardware. Not only has their work been important for Safecast's own documentation, and for updating future radiation sensors, but the process of trying to find the source of anomalous readings was an authentic learning experience in how science is actually done.

> ❝ **I felt I was trying to get my students interested in science from the map I was making, when I should be taking them to the country itself.**

↗ Students measure water samples for pH, temperature, phosphates, nitrates, dissolved oxygen, biological oxygen demand, iron, copper, chlorine, hardness, and coliform bacteria. The data is linked to points on an online map.

What's Next? DIY Sensors

With its large and growing network of citizen scientists, Safecast is an example of where to take a project such as TIGER. Our goal is to involve more schools, students, and volunteers and expand our geographic coverage. However, this growth will likely run into some bottlenecks very quickly.

While our current equipment costs, primarily for water quality testing kits, are low, they involve the use of chemical indicator tablets which can only be used a half-dozen times before being depleted. In addition, these kits limit the types of data we can collect.

The solution, in true maker fashion, is to build our own environmental sensors. Developing cheap environmental sensors has now become possible as the cost per sensor has dropped to near $1, and standard proces-

sor platforms such as Arduino have become readily available. There are a number of immediate benefits to going this route. First, electronic sensors can be used thousands of times, removing a limit on the amount of data collected as well as the cost.

Second, developing our own sensors gives us far greater flexibility in the type of data collected. Given commercially available technology, we can readily monitor everything from carbon monoxide levels to soil salinity to ultraviolet radiation.

Cellphones as Sensor Network Nodes

While tens of dollars per sensor system does represent a significant cost reduction, there is an even cheaper method for developing a citizen science project such as TIGER: cellphones. Not only have cellphones become a globally ubiquitous technology, but they also contain an increasingly complex set of processors and sensors.

What students in TIGER can do, as a number of other developers have done, is create apps to harvest data from such sensors as the GPS unit and camera in order to record everything from the geographic distribution of invasive species to the amount of atmospheric haze.

Make Your Own Science

TIGER and similar projects give students the opportunity to learn analytical and reasoning skills by going into the field to collect and analyze data for their own research projects.

The real excitement, though, will start when students across the world begin to pool their work across different schools and lab groups, design and build their own equipment, and modify their own devices. In short, the future of science will come to those who make their own science. ◪

Ariel Levi Simons is a physics and research projects teacher at Wildwood School in West Los Angeles. He has a background in high-energy particle physics research, and is currently working on building a partnership between high schools and established research institutions.

UNBLENDERS, DREMELFUGES, AND OPTICAL TWEEZERS

Making research-grade equipment from repurposed parts.

BY LINA NILSSON

Two of the most useful DIY projects I know of are a modified record player that doesn't make a sound and a kitchen blender that unmixes its contents. Both are research-grade science equipment. Scientists, both amateur and professional, build equipment to do real and important experiments. The instruments range from simple modifications of $10 gadgets to $100,000 precision instruments built from scratch. DIY equipment is an alternative that can save money and be easier to repair than closed-box commercial versions of standard equipment. But at the same time, some of the most cutting-edge research done today is possible only through novel DIY instruments.

Standard methods for identifying proteins and DNA with dyes require an overnight destaining step on a sample **rotator**. Growing a variety of single-celled organisms is done with similar agitation. Rather than buying a commercial rotator for $500, either procedure can be completed by modifying a gramophone to spin sample containers rather than LPs. By adding an adapter to hold test tubes, your kitchen blender can become a centrifuge for separating samples such as bacteria cultures or blood components.

The **centrifuge** is a workhorse of the research biology lab, and there are many research-grade DIY options. One example is a small centrifuge that requires no more than 30 minutes of simple modifications to a handheld eggbeater. Doesn't sound like a research-grade instrument? Think again. It was designed by one of the world's most well-known chemists, George Whitesides, at Harvard. In order to use such simple DIY equipment — or any equipment — with controlled research studies, the performance has to be quantified and documented. Whitesides' eggbeater design has been published with both mathematical calculations and experimental performance tests.

Moving out of the kitchen and into the garage for inspiration, we get higher performance (faster spinning) with Cathal Garvey's "Dremelfuge," a 3D-printed centrifuge rotor attached to the popular handheld rotary tool. Garvey has used the Dremelfuge attached to a standard drill for small-scale isolation of plasmid DNA from bacteria, an important step in many studies of genes and proteins.

To sterilize tools before experiments, scientists use an **autoclave**, an instrument that is essentially a large and sophisticated pressure cooker. For many applications, borrowing the latter from your kitchen will work just as well, assuming you can work with smaller batches and are willing to wait a little longer (pressure cookers operate at a lower pressure than autoclaves and for obvious reasons also are not designed with a built-in drying cycle at the end). Just as for a commercial autoclave, you can test that the pressure cooker did its job by adding autoclave tape on whatever you're sterilizing — if black stripes appear, the temperature got high enough (typically 121°C).

Keeping things sterile and preventing contamination is important for biological experiments. For this reason — and also because nobody wants to continuously stir a liquid for hours on end — a laboratory device called a **magnetic stirrer** is commonly used to mix and prepare liquids. The stirrer platform relies on a rotating magnetic field to continuously and evenly spin a small magnetic bar that is put inside the liquid container. A basic commercial version costs $250–$1,000, but university researcher Malcolm Watts has built a DIY version for well under $30 that is so elegant my university colleagues don't even realize it's home-built (see teklalabs.org for the design). Unlike any commercial version I know of, Watts' magnetic stirrer runs off of a battery, so I can easily move it around the lab as needed and researchers can use it in remote field locations or developing countries without access to wall power.

In addition to the flexibility and affordability of DIY, researchers make their own instrumentation to be able to make repairs in-house. Indeed, not too long ago, in-house instrumentation repair and design was common at research institutions, but you'll be hard-pressed to find an equipment designer or scientific glassblower at my work today. Brian Millier, an instrumentation engineer at Dalhousie University, says, "30 years ago I performed about 95% of all of the repairs needed on our commercial instruments." Today, commercial equipment is more complex, and parts are miniaturized and not readily sourced or replaced. The instrumentation

Bertram Koelsch

> ❝ **Some of the most cutting-edge research done today is possible only through novel DIY instruments.**

⚊ **FABBED LAB** The DIY science group Tekla Labs, founded by the author, creates free instructions on making "standard laboratory equipment using locally available supplies."

is simply not meant to be repaired in-house. Millier claims, "I can now perform less than 50% of the required repairs, even though I have 30-plus years' experience in this field."

With less work on commercial equipment, Millier has turned to making his own equipment for university researchers and teaching labs. For example, Millier has built a $150 **photometer** that uses an RGB LED for the light source and a low-cost RGB light sensor to detect colors in experimental solutions. In the lab, one of the many uses of photometers is the detection and quantification of proteins that have been tagged with fluorescence

markers (see Nobel Prize in Chemistry 2008).

While commercial spectrophotometers, which cost thousands of dollars, have full spectrum capabilities, Millier's version measures absorption only at three distinct wavelengths, corresponding to green, red, and blue. Indeed it is common for DIY equipment to have more restrictive functionality than their commercial counterparts, but for many routine assays, this core functionality is sufficient. For protein tagging, most researchers

sophisticated instruments that use laser light to very precisely move small particles, such as proteins, within a three-dimensional space. There are currently a handful of commercial options, but not only are they more expensive, they are not easily modified and will in general have inferior experimental performance. So, the vast majority of researchers build their own. Indeed, if you have $100,000 for optical parts and a few months of free time, you can build your own **optical tweezer** setup to

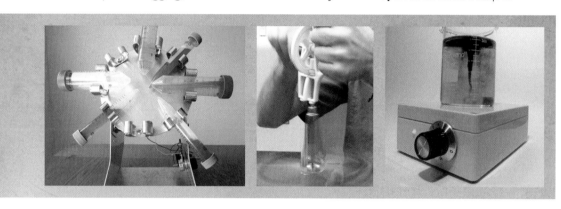

✎ (Left to right) This sample rotator uses a DC motor to provide continuous mixing of lab samples. Centrifuge made from a hand-held eggbeater. This $30 DIY magnetic stirrer does the job of its $250–$1,000 commercial counterpart.

focus on the limited set of green, red, and blue wavelengths of common tags.

Instruments used to copy DNA start at around $5,000 and can get significantly more expensive. While these PCR machines (see Nobel Prize in Chemistry 1993) can do other fancy tricks, their key function is to cycle the sample temperature. Millier built a **PCR machine** for researchers at his university using a custom controller and a toaster oven. For the ultimate low-tech DIY, the original method of manually moving your sample between tubs of temperature-controlled water can also give you the same result.

Not all DIY equipment solutions are low-tech. Many high-end commercial instruments started out as projects by single scientists and engineers and were later commercialized. This driving force of building-it-yourself remains today. For example, optical tweezers are

mechanically manipulate proteins with sub-nanometer (10^{-9} meter) and 0.1-second resolution. Commercial designs will catch up and perhaps eventually surpass these DIY designs, as they did for other high-end instruments such as electron microscopes. But at least for now, DIY reigns supreme.

I regularly meet scientists who have designed their own instruments, from eggbeater centrifuges to high-precision optical tweezers. However, unlike the many other DIY forums, research-grade science equipment does not have an active community for sharing innovations. If we're serious about open science, we need to not just change how we share results but also ease access to laboratory infrastructure and experimental inputs. We need to start sharing our equipment designs. Help me at teklalabs.org. ◪

Lina Nilsson is a scientist and traveler who dreams of new ways to improve research resources globally. She is the founder of the DIY science group Tekla Labs (teklalabs.org) and is currently a biomedical researcher at UC Berkeley.

Lina Nilsson

LORD KELVIN'S THUNDERSTORM

Create high-voltage sparks using nothing more than falling water as the source of energy. **BY MATTHEW GRYCZAN**

With a few household items and a trip to the hardware store, you can whip up one of the all-time favorite projects of science experimenters: Lord Kelvin's Thunderstorm, a high-voltage electric generator that uses nothing but dripping water as its source of energy.

It's cheap to build, and perfect for anyone who loves to experiment because it can be made from a wide variety of materials once the basic configuration is understood.

Most importantly, you have a chance to contribute to the understanding of this fascinating effect and improve on technology that one day may spawn alternative energy sources, such as wind generators that don't have moving blades or rotors. Building Lord Kelvin's generator doesn't take a big research budget, and its "open source" design plays into the hands of makers with ingenuity.

How It Works

Perhaps best known as Lord Kelvin for his scale of absolute temperature, Sir William Thomson was an Irish-Scottish physicist and mathematician who in 1867 invented what he called his "water-dropping condenser." Kelvin's electrostatic contraption generates voltage differences from falling streams of water, similar to the way charged water droplets in a thundercloud generate the static discharges we see as lightning.

The basic setup is 2 streams of water that flow through 2 hollow electrical inductors,

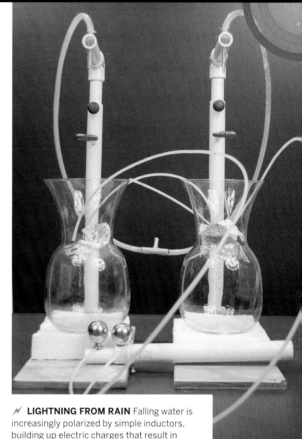

⚡ **LIGHTNING FROM RAIN** Falling water is increasingly polarized by simple inductors, building up electric charges that result in shooting sparks. All the work is done by gravity.

and 2 catch basins that capture the falling water. Each basin is connected to the inductor on the opposite water stream (Figure A, following page).

With gravity as the energy source, the water drops carry electric charges down to the basins, where the electric potential continues to rise until either a spark of electricity jumps across an air gap or it leaks away unnoticed at any sharp edges of the device. My favorite explanation of the phenomenon is by MIT physics professor Walter

Matthew Gryczan

A Most Curious Apparatus (Make 2)

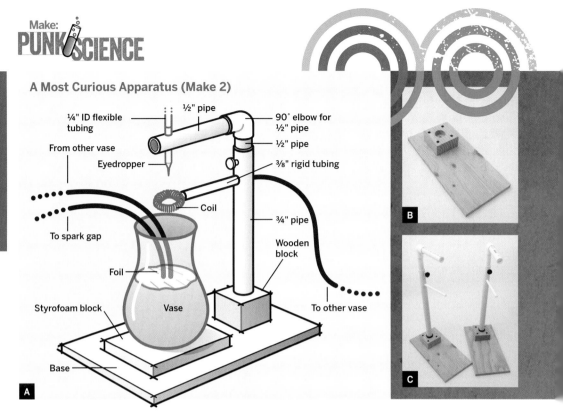

Labels in Figure A:
- ¼" ID flexible tubing
- From other vase
- Eyedropper
- ½" pipe
- 90° elbow for ½" pipe
- ½" pipe
- ⅜" rigid tubing
- Coil
- ¾" pipe
- To spark gap
- Foil
- Wooden block
- Styrofoam block
- Vase
- To other vase
- Base

A **B** **C**

Lewin (see Resources). It goes like this:

Let's say inductor A, by chance, has a slight positive charge. The inductor polarizes the water falling through it, giving each drop a negative charge. These drops fall into basin A, charging it negatively. Now it gets interesting: basin A, remember, is connected to inductor B, giving it a negative charge as well.

Once negatively charged, inductor B polarizes its water stream too, giving its drops a positive charge. These positive drops fall into basin B, charging it positively, and basin B in turn adds its positive charge back to inductor A. The cycle becomes a runaway positive feedback loop, increasing the charges in both basins exponentially until the potential difference is so great (up to 20kV) that a spark jumps the spark gap between basins.

Given its simplicity, Lord Kelvin's Thunderstorm can be made from a myriad of materials. A quick survey of YouTube shows generators made from styrofoam plates, soda straws, aluminum foil, and soup cans.

My design takes a little more time and care to build than one fashioned from soda straws, but the maker is rewarded with a generator that can be used to conduct reproducible,

measurable experiments, including a remarkable phenomenon where water drops orbit the inductor like tiny satellites.

START
1. Build the bases.

You can substitute a number of materials here — thicker plywood for the stands, different diameter plastic tubing, or plastic basins rather than glass — so long as they fit together properly. But keep in mind that the better you insulate one inductor/basin set from the other, the better your generator will work.

I made 2 identical stands in the following way. The base of each stand is a 6"×12" piece of ½" plywood with a 1½" cube of scrap 2×4 glued on top, centered 2½" from one end. Find the center of the block and drill a 1" hole with a spade bit or hole saw, ½" to ¾" deep (Figure B).

I recommend putting a water-repellent finish on the base: water inevitably falls on the base and soaks into the wood if it's not protected, causing possible leaks of high-voltage electricity. You could also substitute a plastic base.

MATERIALS

Plywood, ½" or thicker, 6"×12" pieces (2)
Scrap wood, 1½" cubes (2)
Water-repellent finish for wood (optional)
but recommended
PVC pipe, ¾", Schedule 20: 14" lengths (2) and 3" length (1) for the stand columns and the spark gap body. Schedule 20 pipe has thin walls, so a standard ½" pipe will fit inside.
PVC pipe, ½": 5" lengths (2), 3½" lengths (2), and 6" length (1) for the dropper arms and the spark gap body
PVC pipe fittings, ½", 90° elbows (2)
Thumbscrews (3) or threaded knobs or machine screws, whatever you have handy; for setscrews on the columns and the spark gap body
Plastic tubing, rigid, ⅜" OD, 6" lengths (2) for the inductor arms
Eyedroppers, glass, standard size (2) These are about ⁹⁄₃₂" in diameter.
Plastic tubing, flexible, ¼" ID, 5'–10' length for water hoses, depending on how far away your water source is mounted
Tee connector, ¼" ID to fit the flexible tubing. I used a brass tee.
Bucket or faucet or other water source
Scrap wire to clamp water hoses to eyedroppers
Coaxial cable, insulated, 6' lengths (2) the kind used to hook up cable TV
Copper wire, 14 gauge, bare, 32" lengths (2) I stripped the insulation off 14-gauge home wiring.
Bullet connectors, 16–14AWG, 0.156": male (2) and female (2) (optional) These quick-disconnect terminals are handy for experimenting with inductors.
Aluminum foil, 6" square pieces (2)
Glass or plastic basins (2) I used a couple of glass flower vases. Glass jars work well.
Scrap styrofoam blocks, or 3"–4" PVC pipe scraps (2) to insulate your basins from the bases
Round brass furniture knobs (2) or other round metal objects for the spark gap, such as metal beads, hollow metal balls, etc.

TOOLS

Saw
Drill and bits: 1" spade bit or hole saw, ⅜", ⁹⁄₃₂", and bits to match your screws e.g., #7 for ¼-20
Screwdriver to match your screws
Screwdriver, Phillips head for winding inductor coils
Hot glue gun
Box cutter or similar knife
Wire cutters
Wire strippers

2. Build the stands.

The main column of each stand is a 14" length of ¾"-diameter PVC plastic pipe. About ½" from one end, drill a hole slightly smaller than whatever machine screw you have handy (I used a ¼-20 NC thumbscrew and #7 bit) and thread a screw into the hole, either with a tap or with the screw itself (the plastic is soft enough for you to make your own threads). This screw acts as a setscrew to hold a smaller piece of pipe inside the column in place when you adjust the position of the water droppers.

Make the arm for each inductor from a 6" length of ⅜" OD rigid plastic tubing. To hold the inductor arm, drill a ⅜" hole completely through the ¾" column pipe, about 2½" from the same end where you put the screw.

Now make the arms to hold the water droppers. Each arm consists of 5" and 3½" lengths of standard ½" PVC pipe, held together with a 90° elbow fitting (Figure C).

If you're using 2 standard glass eyedroppers like I did, drill a ⁹⁄₃₂"-diameter hole about ¾" from one end of the 5" pipe. You may have to expand the hole slightly so the eyedropper slides in easily.

3. Make the inductors.

Use a box cutter to strip all the outside insulation and shielding off the coax cables, to get at the central strand of insulated wire. (You can use standard insulated wire, but I believe the central strand of coax cable carries the high-voltage charges better without leaking electricity.) Now strip 1" of insulation from each end of the cables, and about 8" of insulation from the middle of each cable to expose the bare wire.

I made 2 inductors, from coils of bare 14-gauge solid copper wire used for home wiring. Cut a 32" length of wire and wind it like a spring around a Phillips screwdriver, leaving 1½" leads on each end, then slip the coil off the screwdriver. Trim the ends so that both leads are on the same side, then gently bend the coil into a donut, putting the 2 leads together (Figures D and E, following page).

D

E

F

G

H

I

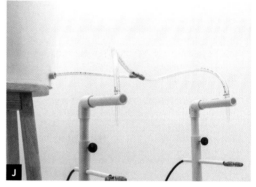

J

You can solder the 2 leads of the inductor directly to the end of its cable, but you'll need to melt and resolder this connection if you want to experiment with different inductors, such as copper tubing or flat coils. If you want to switch inductors easily, solder a 0.156" male bullet connector to the leads of the inductor, and a female bullet connector to the lead of the cable (Figures F and G). These quick disconnects make it easy to experiment to see different inductors affect the generator's operation.

4. Assemble the generator.
Assemble the 2 stands, string the cables though the inductor arms, and use hot glue to attach the inductors to the arms so the inductors are rigid.

Coil the bared wire in the middle of each cable, and fold a 6" square of aluminum foil around each coil (Figures H and I). Wedge or fasten each foil plate to a glass or plastic basin so that the water will drip onto the foil, and so that the left inductor is connected

to the right basin, and the right inductor to the left basin. Whatever you use for basins, I recommend insulating them from the bases with a styrofoam block or large PVC pipe scrap.

Assemble the remaining parts as shown in the diagram, and connect the plastic tubing to the eyedroppers from either a faucet or a water tank, such as a bucket (Figure J).

Then adjust all the components so water drips through the inductors, and soon you'll get sparks jumping between the 2 free ends of the cables.

5. Make the spark gap.
I highly recommend making a spark gap because it allows you to see the electrical discharges easily, to adjust and measure the voltage that the device generates, and to vary the power and frequency of sparks.

The body of the spark gap is a section of ¾" Schedule 20 pipe, with a section of ½" plastic pipe inside, and a setscrew through the outer pipe. Install this screw the way you did the dropper arm screws (Figure K).

K

L

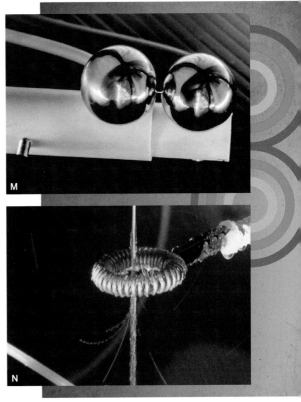

M

N

Connect the free ends of the cables to your round metal terminals. Mount one terminal to the end of each pipe, then slip the smaller pipe into the larger. Use the screw to adjust the distance between the 2 terminals (Figure L).

6. Experiment!

Your generator will work differently if you change the speed of the water flows, the design of the inductors, and the length of the spark gap. Or try varying the distance of the eyedroppers from the inductors, or the types of fluid (would an acidic liquid work?). Even the orifices of the eyedroppers have an impact on the generator, because of drop size.

This project is sensitive to humidity so keep your surroundings as dry as possible. Use ordinary tap water; it has plenty of the ions needed to begin the charging cycle. And lastly it's absolutely critical that there are no sharp points anywhere in the circuit.

» Shooting Sparks

You'll quickly learn when the generator is ready to spark because the water flow changes as the voltage builds up. The water stream breaks into drops and is often deflected as it passes through the inductors.

Vary the spark gap to get many smaller sparks or fewer large sparks (Figure M).

» Orbiting Droplets

To see water droplets orbit the inductor, open the spark gap completely and let the charge build up in the inductor as high as it can. Depending on the light, look at the inductor

from different angles until you can see tiny water droplets spinning around the inductor rather than falling into the basin (Figure N).

» Neon Flash

A neon test lamp, such as those used to check whether an electrical socket in the home is wired correctly, will flash if you touch both sides to the wires of the Kelvin generator. ◢

Resources

- ◼ The author's Kelvin water dropper in action: makeprojects.com/v/31
- ◼ MIT physics professor Walter Lewin explains the Kelvin water dropper: video lectures.net/mit802s02_lewin_lec15 (starting at 27:20)
- » Thunder, lightning, and the Kelvin water dropper: makezine.com/go/kelvin

Matthew Gryczan is a former manufacturing engineer and newspaper reporter who has been a lifelong basement tinkerer. In his day job, he writes news releases for science and technology companies at SciTech Communications (scitechcommunications.net).

Matthew Gryczan (N)

BUILDING LIVING MACHINES

The birth of a kitchen-table biotech company.

BY JAMES PEYER

I've always been amazed by manufacturing. A Ford automobile factory in Kansas, one of the biggest in the world, makes a car every minute. That's 1,440 cars every day. But factories come in all shapes and sizes; they can even be microscopic. A single bacterium can be turned into a factory that makes proteins, the building blocks that make all living things work.

A single bacterial factory can make 500 to 600 proteins every second (43 million each day), all the while dividing about every half hour to generate over 200 *trillion* identical factories in 24 hours, each one ready to make another 200 trillion tomorrow. Given enough space and food, one engineered bacterium will generate a limitless number of proteins.

Who wouldn't get excited about the possibilities of biotechnology?

Last July, Genotyp (speakscience.org) was in Dearborn, Mich., at the Henry Ford Museum for Maker Faire Detroit, 2011. We brought along some genetically engineered bacterial factories happily generating a protein from jellyfish that gives off a green glow whenever it's hit with UV light. One question kept popping up from our fellow makers: "When can I do that?"

You can start doing genetic engineering now! A few hundred dollars will get you all the reagents you need: with some baths of hot water you can perform polymerase chain reaction (PCR), and some simple chemicals will allow you to do bacterial transformation, all the techniques required to do cloning. The difficulty with amateur biology isn't the availability of tools; the hard part is learning how to design and perform biotech experiments.

Learning Biotech

Imagine yourself dropped into a state-of-the-art university lab. You've got a million dollars of machines and biotech tools at your disposal. Anything you come up with is possible in this kind of environment, but where do you begin? To start your project, you're going to need a good grasp of the biotech workflow.

Being able to make a step-by-step plan for accomplishing a project is critical in any discipline. Before discovering the MAKE community, I had no idea how to make a working circuit or program basic computer instructions. To learn these skills, I started with very simple tutorials: wiring up a circuit to blink an LED and writing a program that would happily chirp "Hello, World!" on my computer screen.

These baby steps familiarized me with the workflow of putting together circuits and writing code. There were good online instructions because thousands had done the exact same thing before me. After completing the tutorials, I knew what tools I'd be working with and how to access them. Then I could start thinking about my own projects within the scope of a technological workflow.

Right now, in order to learn biotech, you have to become an apprentice. That means working in a lab, most likely at a research university. The apprentice model is wonderful for achieving mastery in a discipline, but lacks

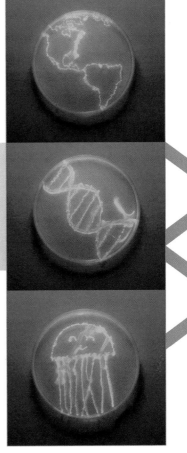

" Even middle school students can clone genes.

scalability. Since anyone who wants to learn has to find a college professor to teach them, very few people learn biotech. Compare that to a one-dollar LED kit or a free, downloadable programming tutorial. It's no wonder we have a lot more programmers and engineers than we do biotechnologists.

Genotyp was founded to help with this problem. We wanted to make getting started with biotech fun and easy, so we created "Cloning a Fluorescent Gene." We distilled the biotech workflow to its minimum components while still providing a clear picture of genetic engineering from start to finish. It's a great tool to teach biotech to high school students. Other efforts, like the BioBrick foundation and the iGEM competition are encouraging students to use biotech to create machines out of living cells. Resources and technologies to make biotech accessible to beginners and amateurs are slowly taking shape, making it easier to speak the language of science.

Making with Biotech

Understanding how life uses DNA and protein to create complex machines is an incredible challenge. Our genomes contain over four billion DNA letters, a language that we don't speak or understand completely. To figure out the language of life, we have to study DNA in pieces, taking out a little bit at a time and figuring out what each piece does in the whole system. This task is like being handed a hard drive filled with billions of 1s and 0s and having to figure out how binary code makes

all the parts of an operating system.

Progress may be slow, but in the last 20 years, biotech has started showing some incredible results. Scientists have created new tools to study and modify DNA so that even middle school students can clone genes. We know almost all the genes in the human genome and even have pretty good ideas of what a lot of them do. We know even more about the genomes of simpler organisms like yeast and bacteria.

Maybe in a few years, next to the Ford plant, there will be another factory, one that employs trillions of microscopic workers given very detailed DNA instructions. Using only compost for food, they'll work day and night to generate a sustainable source of gasoline needed to run the new Fords next door. ☑

James Peyer is the CEO and co-founder of Genotyp (speak science.org), a small Michigan-based biotech company dedicated to giving high school and college students an introduction to modern biology research.

Kyle Lawson

DRIVE-BY SCIENCE

The evolution of Safecast's bGeigie radiation logger.

BY SEAN BONNER

In the year between March 2011 and 2012, a gang of hackers collected and published more precise and useful data about radiation contamination in Japan than any other organization, including TEPCO and the Japanese government combined. And we did it almost entirely with devices we soldered, bolted, and duct taped together at the Tokyo Hackerspace. We call ourselves Safecast.

Let's back up for a second, to the days following the Tōhoku earthquake and resulting tsunami, when it was becoming clear that things were going … well, less than awesomely at the Fukushima Daiichi Nuclear Plant. Everyone was looking for details about what was actually happening, or what had actually happened, or what might happen. And they weren't having much luck. I know because I was one of those people, and I spent countless hours trying to find any useful information, to no avail. I was asking friends, friends were asking me, and before too long, we had a whole group of people asking the same questions and trying to find answers. Finally it became clear that the only way we were going to get any useful data was to go out and collect it ourselves.

So what's the best way for a handful of people to collect data about nuclear fallout around Japan? Huddled in an office in Tokyo, we wrestled with the question, and then Ray Ozzie (former chief technical officer at Microsoft) delivered the answer at a brainstorming session with other Safecast members: strap a Geiger counter to a car and go for a drive, obviously.

Our first shot at mobile radiation logging was as hands-on and basic as you could possibly get. A minimum viable product if there ever was one. Safecast volunteer Dave Kell had the (mis)fortune of doing the very first measurement drive with a system we'd rigged up for him. Actually, the word "system" is a bit generous — we gave him a Geiger counter and an iPhone with a preinstalled Flickr account and instructions to "drive through Fukushima and every once in a while stop and take a photo of the reading and upload the image to Flickr."

The photo would be geotagged to give us a location, and the reading shown on the display in the photo provided the data. These of course had to be entered by hand. We also had to note variables such as whether the reading was taken inside the car or out, in the air or on the ground, etc. Despite all this, it actually worked and gave us our first real data set to analyze.

But if you're thinking the system could use some refinement, you're right. A few others used this method for some drives in the following days, but we knew it needed an iteration, and quick.

The first major improvement came from Kei Uehara and some of his students from Keio University. Like all first major improvements, this was enabled with duct tape. The "upload-a-geotagged-image-to-Flickr-and-input-data-by-hand" method was still in effect, but variables were reduced by duct taping the Geiger counter to the window of the car and taking a photo every five minutes, on schedule. This gave us a more consistent measurement and a clearer route, but was still a pain in the ass. And a pain in the ass drives innovation as well as anything — sometimes better.

While those "by hand" drives were happening, a core group at Tokyo Hackerspace had

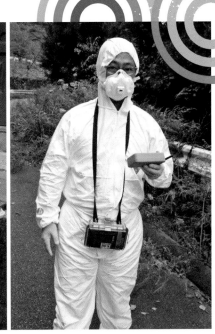

/ **RADIATION SCOUTS** (Clockwise from top left) Two of the original bGeigies outside Tokyo Hackerspace; Dr. Akira Sugiyama of Keio University, walking into the exclusion zone; the Safecast car equipped with two bGeigies; Safecasters (from left to right) Jun Nakamura, Akiba, Pieter Franken, and Robin Scheibler.

Pieter Franken

busted out the soldering irons to work out a more automated solution. Robin Scheibler, Steve Christie, Mauricio Cordero, Akiba, and Pieter Franken spent most waking hours of a solid week drilling, bolting, inhaling solder fumes, and staring at lines of code to give birth to the bGeigie. The name was chosen because the little Pelican case everything is crammed into looks like a bento box — a traditional Japanese lunchbox. This would become a line of devices that are the epi-

center of Safecast data collection. This first bGeigie Classic (as we now refer to it) was a Geiger counter (we used the Inspector Alert by Medcom which has a 2" pancake sensor) inside a modified Pelican case (modified to let radiation in, but keep weather out) along with an Arduino Uno and a GPS module attached to the outside of a car. A line out from the case ran inside the car to a netbook which logged a data point every five seconds, with the added benefit of informing the people

inside what the radiation levels were outside the car. Now this was some serious progress!

The next two devices were variations on this Classic design, but used a Freakduino instead of an Arduino because we thought the onboard wireless might be useful and because Akiba had a bunch of those lying around. One of these was hooked up to a MacBook Air (making it the most expensive bGeigie version to date) and the other to a Dell laptop. The Dell version is now permanently installed in the Safecast car in Japan. While this was a major step forward from the "by hand" method, it was still rather bulky and cumbersome. That was the next issue we tried to solve.

The fourth device we made, which we called the bGeigie Mini, has become the base model for all future devices. It had the advantage of scrapping the laptop altogether. Akiba designed an SD card shield for his Freakduino boards, and Robin wrote the software to power it. With the addition of two AA batteries for power, this became the first self-contained version of the bGeigie. No external power, no cables, no laptops. This was a giant leap for us, but not even a step forward for mankind. We think of this one as something of a working prototype, since we changed a few things right away.

The fifth device swapped out the two AA batteries for two C batteries and modified the Freakduino to run on 3 volts. This was finally a very solid and repeatable design, and the next 20 or so devices we made were modeled on this one. We made minor revisions along the way, trying a different kind of foam to hold the Geiger counter in place, different positioning of the GPS unit, arguing over the merits of using a toggle or rocker switch for the power, different color cases — but essentially the guts of the device stayed the same. The onboard SD card logged the data, but still required someone to manually pull the card out and get the log file, then upload it to the Safecast server. Until early 2012, the vast majority of the Safecast data was collected with one of these models.

Somewhere along the way we realized two

things: 1) While having a laptop in the car was annoying from a logistical and power supply standpoint, it had been useful for the team in the car to have the information it displayed; and 2) The Freakduino boards we were using had built-in wireless connectivity we could be using. This led Robin to build a companion device we lovingly refer to as the bGeigie Ninja. It's a small wireless display monitor that sits inside the car and networks with the bGeigie outside of the car, to give immediate feedback on the outside readings.

We rocked this setup for months before the next major revision. In November of 2011, Joe Moross debuted the bGeigie Pro, which was immediately renamed the bGeigie Plus for reasons I'm forgetting at the moment.

Its awesomeness came from being a truly self-contained device that never needed to be opened, and in fact had a sealed case to keep out water and volunteers' fingers. The added USB port powered rechargeable batteries and made data output a snap. After considerable deliberation, the signature duct tape window was removed (much to Joe's delight and Pieter's disapproval).

With an inventory of bGiegies, combined with an ever-growing roster of people jumping in to lend a hand, as of March 2012 we've collected and published close to 3 million data points. Compare that against the few thousand data points provided by official sources, who I'm going to assume have more resources at their disposal than some guys at a hackerspace.

I really love this story and am so psyched to have played a part in it, because in many ways it's the quintessential example of hacker/punk motivation and DIY execution. Permission wasn't asked. Rules weren't considered. A need was determined and immediately acted on. No one cared that the devices weren't perfect on day one, and they didn't care that they'd be revised on day two. There are any number of schools of thought that would have dictated consulting local authorities, beta testing hardware, and refining and fine-tuning everything before ever heading out into the world. There's certainly a place for that way of thinking, but it wouldn't have produced anything close to the results we got in such a short time frame, and getting data fast was the problem that needed to be solved.

While year one of Safecast clearly was short on sleep, this isn't the end. There's more data to collect, and with the combo of our open designs and committed hardware partners, we hope to be able to reach corners of the Earth we haven't even considered yet.

If this sounds interesting, please check out safecast.org. We'd love to have you get involved too! ◩

When not playing with Geiger counters, Sean Bonner can likely be found at a coffee shop, hackerspace, or Family Mart near you. Find out less about him at seanbonner.com.

◸ Safecaster Dave Kell with the Geiger counter and iPhone he used on the very first mobile measurement run, pre-bGeigie.

Safecast's Data Is for Everyone

At its core, Safecast is all about the data, which is why we've chosen to publish every bit of data we collect under a CCO designation (creativecommons.org/publicdomain/zero/1.0).

CCO is similar to public domain and waives all rights so there's completely unrestricted access and open use of the data allowed to anyone. Data is only as valuable as its usage, and we didn't want any limitations on who could use it or how it was used. We don't even require people to attribute anything to us.

This allows our data to be included in other studies, analysis, services, and large data aggregation projects that would fail if they had to document the hundreds of thousands of sources for each and every data point. Attribution can sometimes be more work than the study itself, and we didn't want that to stand in the way of anyone doing good things with our data.

We already know of scientists contrasting our data against a study of stress levels in certain areas, medical professionals looking for similarities with their own symptom maps, and mega-trend research which may help us better understand what the future holds. We think the more data humanity as a whole has access to, the better off we'll all be.

❝ It's the quintessential example of hacker/punk motivation and DIY execution. Permission wasn't asked. Rules weren't considered.

MY SCANNING ELECTRON MICROSCOPE

BY BEN KRASNOW

Cody Pickens

I decided to design and build a scanning electron microscope (SEM) in my home workshop to see if it was even possible. Spoiler alert: it is. I didn't originally intend to create an SEM that could compare to a $75,000 entry-level commercial model, but the project turned out more successful than I expected. It produces clear, accurate images, and after some improvements that I'm currently working on, it may be practical for hobbyists to build an SEM that's suitable for scientific research for under $2,000.

How SEMs Work

Ordinary optical microscopes shine visible light onto or through a specimen and use lenses to create a magnified image. This works well for many applications, but light can only resolve features greater than about 200 nanometers for visible light. This is small, but not small enough for looking at many interesting biological and material structures. You can use light with a shorter wavelength (i.e. ultraviolet light) to obtain slightly better resolution, but this adds a lot of expense and difficulty for only incremental improvement.

Electron microscopes, in contrast, offer a tremendous improvement in resolution. Like photons, electrons have both particle and wave-like properties, but the wavelength of a fast-moving electron is substantially shorter than that of visible light.

The SEM scans a tiny beam of electrons across a sample, following a raster pattern, and measures the amount of electrons that bounce off each point and onto a nearby detector. For example, if the beam hits a hole in the specimen, the electrons may become trapped and won't reach the detector, but if the beam strikes a protrusion on the surface, many electrons will reach the detector since the protrusion provides more surface area than surrounding flat areas.

In this way, the SEM builds up its image pixel by pixel, and the device's maximum resolution is determined by 2 attributes of the electron beam: its spot size and its scan rate. A smaller spot size will resolve greater detail, and slower scanning improves resolution by raising the signal-to-noise ratio at each spot. So that the electrons are not absorbed, the sample must be conductive or else coated with a thin layer of metal.

This method lets you image 3D objects over a wide range of magnifications without slicing them to bits, and the images it creates look like black and white photographs with a high depth of field. These attractive image qualities make SEMs very common for studying small 3D objects, and they influenced my choice to build this type of electron microscope.

Create a Vacuum

One challenge of SEMs is that the electron beam and the specimen must be manipulated within a vacuum. If the electron beam were fired through air, the electrons would strike gas molecules and scatter, blurring and destroying any image. For the electrons to travel unimpeded from source to sample and from sample to detector, you need a vacuum about a million times lower than atmospheric pressure, or 0.00076 torr, where a torr is the unit of pressure required to support a column of mercury 1mm high. Atmospheric pressure is about 760 torr at sea level.

You can get to these low pressures a few different ways, but my favorite (the least expensive) is by combining a mechanical rotary pump and a diffusion pump, plumbing

Baffles **Water cooling** **Rotary pump**

Diffusion pump

Wire-reinforced tubing

them in series. The mechanical pump reduces the pressure by about 4 orders of decimal magnitude, and the diffusion pump takes it down another 2. For the rotary pump, I settled on a $150 air conditioner pump from Harbor Freight, and for the diffusion pump I bought an air-cooled 3" Varian pump on eBay for about $200.

Diffusion pumps operate by creating high-speed jets of hot oil vapor that push air molecules out of the vacuum chamber. Inside the pump, an electric element heats silicone oil into vapor. After the droplets are done bumping air out, the pump's cooled walls condense them back into liquid, which drips down to the bottom to be boiled again.

I connected the rotary pump to the diffusion pump with ¾"-ID wire-reinforced tubing from McMaster-Carr, where I got most of the hardware and raw material (Figures A and B). The wire reinforcement prevents the tube from collapsing under vacuum. I also included a tee fitting between the 2 pumps and added a digital vacuum gauge that I bought on eBay for about $100. The gauge reads from 0.001 to 12 torr, and was made for refrigeration technicians to use with a vacuum pump.

I didn't have a commercial vacuum chamber, and I wanted the microscope to work inside a transparent enclosure, since its main purpose would be demonstration. So I used a glass bell jar that I found on eBay a while ago. The glass thickness indicated that the jar was built for vacuum use, rather than just for ornamentation or dust shielding. For a base, I used a 1"-thick aluminum plate. I cut

Ben Krasnow

C

a hole in the plate to fit the diffusion pump and machined a water-cooled baffle to go between the pump and plate (Figure C).

The baffle prevents diffusion pump oil from migrating into the bell jar. Boiling oil gets messy, and getting even small amounts of oil into the sensitive parts of a SEM would cause many problems. Air molecules can pass through the baffle's tortuous pathway, but hot oil molecules condense on its water-cooled surfaces and drip back down.

I cut another hole in the aluminum base plate and added an additional vacuum monitor called a Penning gauge, also bought on eBay for about $250. This device measures vacuum from 0.001 down to 10^{-8} torr, and will indicate when the diffusion pump has taken the chamber pressure down to the range necessary for SEM operation.

The first time I pumped down the jar, I started the rotary pump, then exited the garage and shut the door behind me. If the jar imploded, I would be far enough away to escape the wreckage. But below a pressure of 0.01 torr, variations in pressure don't much affect the strength needed for a vacuum chamber. This is a key point that often tricks people. Once you remove 99% of the air molecules, there are so few left that they

exert almost no pressure on the inside wall. Removing more doesn't change much. If a container can withstand 10^{-1} torr, it can probably safely hold 10^{-11}.

Spark Plug Power

Ordinary automotive spark plugs are designed to supply insulated high voltages through metal walls and across pressure differentials, so I used them to bring power for the electron gun into the SEM chamber.

I drilled and tapped a series of holes in the base plate to hold the spark plugs, and added O-ring glands. I also made some low-voltage pass-through connections for other circuitry using wide-head screws sealed to the plate with Buna-N (nitrile) washers. And to let users move a small stage to locate the specimen under the electron beam, I added spring-loaded teflon shaft seals that transfer rotary motion through the base plate while the chamber is under vacuum.

The Electron Gun

There are many ways of generating electrons for an electron microscope, but the easiest is to simply heat up a piece of wire. This goes by the exciting name of *thermionic emission*, and these filaments are used in vacuum tubes and cathode ray tubes; they make the orange glow inside the back of old TVs and radios. From eBay I bought a set of tungsten filaments with ceramic insulator holders that were originally made for use in commercial SEMs.

I connected the filament to a low-voltage power supply that I built from a Variac variable transformer, isolation transformer, bridge rectifier, and smoothing capacitors. I originally fed low-voltage AC to the filament, but that resulted in image quality issues, so I designed an unregulated but smoothed DC power supply.

Once the filament is glowing, it emits lots of electrons in all directions. To motivate them into a single direction, you need to apply high voltages across pieces of metal strategically arranged around the filament. The whole assembly is termed an *electron gun*, and when the applied voltage is 10kV, my gun

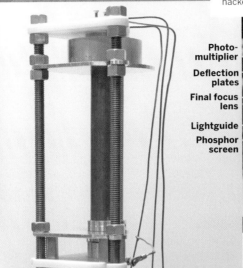

Photo-multiplier
Deflection plates
Final focus lens
Lightguide
Phosphor screen

E

Stage

D

fields that bend and shape the beam something like the way glass lenses bend the paths of photons (Figure D).

Most commercial SEMs use magnetic fields to focus the beam because of their bending power and lower voltage requirements, but I used electric fields, because they don't require custom-machined precision iron pole pieces. I used copper pipe and teflon insulators to construct 2 electrostatic lenses, which are nothing more than 3 lengths of conductive pipe, insulated from each other and arranged inline. As the electrons pass through the charged pipes, their trajectory is affected by the polarity and magnitude of the voltage applied to each. With the correct voltage and geometry, the beam of electrons will focus down to a tight spot on the specimen.

shoots electrons out in a stream at about 2% of the speed of light (6,000,000 meters/second). To supply this voltage, I use a regulated high-voltage supply that I bought at a surplus sale, and I can adjust the voltage to fine-tune the electron velocity.

Focus the Beam

The beam from an electron gun is narrow, but not nearly fine enough for useful electron microscopy. To focus the beam, an SEM needs to run it through electron optics — controlled apertures and electric or magnetic

Scan the Sample

In the first SEMs, the process of scanning the sample and displaying the image were intertwined. The scan beam was steered over the sample in sync with a raster pattern that the CRT beam traced over the screen phosphors, and the microscope's emission detector was used to drive the beam intensity in the CRT.

I took this same approach because of its simplicity; to capture images I currently aim my camera at the screen. But for the next version of my SEM, I am implementing a digital

image storage system that will record the sample's surface emission (image brightness) pixel by pixel.

To perform the synchronized scan and display, I bought 2 identical analog oscilloscopes (eBay again) and took one of them apart. Analog oscilloscopes use oppositely charged pairs of metal plates to deflect the electron beam in their CRTs, with plate size, spacing, and applied voltage determining the amount of deflection. So I removed the CRT from the disassembled scope and rerouted the wires driving its x-axis and y-axis deflection to smaller plates mounted in the SEM column.

To create the horizontal and vertical scan patterns, I built a simple raster generator similar to what's inside a TV, but made from 555 timer chips. I fed its output into both the hacked scope, for steering the SEM beam, and the intact scope set to x-y mode, for driving the display.

Pick Up the Signal

To generate its signal, the SEM detects the quantity of electrons that are emitted from the specimen surface as the electron beam strikes it. But it's a relatively small number within a small range, so it needs to be amplified. (Figure E, preceding page).

To accomplish this, the electrons are attracted toward a phosphor screen which converts them into flashes of light. The flashes of light are then converted back into electrical signals and amplified by a photomultiplier tube, which consists of a photocathode that produces electrons when hit by photons, and a series of 12 dynodes that generates an avalanche of electrons about 10^6 greater in number than the first bunch. The detector is positioned to one side of the stage. It has a curved light guide so that the phosphor screen faces the sample and the photomultiplier runs up vertically.

The signal from the multiplier tube is then fed into the z-axis or blanking input on the intact oscilloscope. At fairly fast scan rates, the oscilloscope will then display an image from the SEM at live video rates.

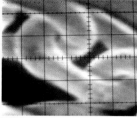
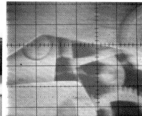

Results

So far, I've just used the SEM to image conductive objects (Figure F), since non-conductive objects must be coated with a vanishingly thin layer of metal before being imaged, done in a sputtering chamber. I may end up building one from a power supply and vacuum chamber.

Biological samples must be dried via special means so that the sample doesn't lose its structure as the water evaporates. You can repeatedly soak the sample in alcohol until the alcohol replaces the sample's internal water almost entirely. Then the sample is placed in a chamber and submerged in liquid CO_2 at about 700psi. Finally, the CO_2 is heated under pressure until it becomes supercritical, a liquid with no surface tension. I've built a supercritical drying chamber and used it to make homemade aerogel.

Meanwhile, I'm also developing a detector system that uses an electron multiplier instead of a photomultiplier, for greater simplicity, purity of the signal path, and to allow the SEM to operate without a light shield (I use heavy black plastic) covering the bell jar, thus improving the signal-noise ratio. ⬈

Ben Krasnow (youtube.com/user/bkraz333) works on top-secret video game hardware at Valve Corporation in Seattle. After work, his projects usually involve home science, circuit design, machining, material selection, and fabrication.

ROBO ROACH

Building your very own cockroach cyborgs.

BY TIMOTHY MARZULLO, PH.D., AND GREGORY GAGE, PH.D.

You wake up with a desire to do some neural engineering. You open up your insect terrarium, bring out your pet flower beetle, your tobacco moth, or your cockroach, and turn on a switch on the insect's back. With your remote control, you begin steering your insect around the room.

Juan Leguizamon

While this may sound like science fiction, remote controlled insects have been around for 15 years. Starting in the late 1990s, two groups of researchers (at the University of Tokyo and the University of Michigan) achieved rudimentary control of cockroach turning. The American work was undertaken with the ultimate goal of creating "hybrid robots" able to do disaster relief monitoring and reconnaissance. The work was published in small trade journals and received some press, but went largely unnoticed. The research eventually faded away, with the scientists working on other adventures.

Years later, two more universities (Cornell and UC Berkeley) achieved notable success and press for their proof-of-concept work in flying beetles and moths using modern small-electrode design and innovative surgical techniques.

Cool? Weird? Revolutionary? Horrendous? It depends on your disposition. Perhaps you're inspired by these insect cyborgs and you want to work on such experiments as well. Unfortunately, unless you're in one of these four research labs (a total of maybe 15 people in the world) you won't have access to the extensive equipment and government funding required to build your robot insect interface.

A DIY Robot Insect Interface

Sound totally unfair? You're in luck. For the past two years, we at Backyard Brains have been working to make neurotechnology previously found only in advanced research labs available to students of all ages (well, mainly over the age of 12). By making equipment that's cheap and easy to use, we're making it possible for everyone to do neuro-experiments and accelerate our understanding of the nervous system.

For our latest project, we wanted to develop a super-low-cost "RoboRoach" to demonstrate principles of neurostimulation to students. This project began as a senior design assignment we managed and sponsored at the University of Michigan in 2010.

For cost reasons, we didn't want to build our own remote control circuits from scratch, so we scoured toy stores for the smallest, lightest, cheapest remote control circuit we could find (a tough job, but someone has to do it). We settled on the Hexbug Inchworm, as the circuit is light (1 gram), runs on only 3 volts (so we can use a small coin cell battery), has 2 degrees of freedom, and costs only $20.

To steer the RoboRoach, we removed the circuit from the toy and fed the motor controller output into the cockroach antennae using very thin silver wires (0.003" in diameter, or 40 gauge). The neural tissue inside the antennae likes biphasic pulses (both + and − square pulses), but the Hexbug powers its motors with DC (direct current). We were able to convert this signal using the most common timer in existence, the 555 (actually, a low-power cousin, the 551). By configuring the timer chip into an astable mode and placing a capacitor on the output (removing DC offset), we were able to make the stimulator circuit deliver 55Hz pulses. This is right in the range that neurons prefer.

The final total package, which includes the connector, 2016 battery, our custom 551 circuit boards, and Hexbug platform, weighs 7 grams. A large cockroach weighing 3 grams can carry upwards of 9 grams for 10-minute experiments. The backpack is detachable so the cockroach only wears it temporarily (when not plugged in, the bug is free to eat fresh lettuce, drink water, make babies, run around in the dark, and do what ever cockroaches do and think about).

How to Steer a Cockroach

To get the cockroach to turn when running, we take advantage of its natural motor behaviors. Cockroaches like to run next to a wall. When their antenna brushes against something, they turn so that they're running parallel to it. This is part of their "escape mechanism." The biphasic pulses trick the bug into thinking it felt a brush on one of its antenna, by stimulating the nerves that normally conduct information about things the antenna touches. The pulses cause the nerves to fire electrical messages (called "spikes") to the brain. The end result is that the cockroach will turn in the opposite direction of the antenna we activate through stimulation.

As biologists, we must emphasize that the cockroaches are not really robots. They have fully functioning nervous systems that show some wonderful properties like learning and adaptation. The reason the research was never ultimately pushed further is that insects eventually learn to ignore the stimulation. Our brains (as do theirs) eventually disregard things that have no behavioral relevance (things that neither hurt nor help us). After approximately 10 stimulation trials, the cockroach turning response noticeably diminishes, and eventually stops all together.

But — if the cockroach is placed back in his cage for a couple hours, it forgets and the stimulation works again. After about a week, however, the stimulation stops working entirely. We don't know why, but we suspect it's most likely due to biofouling of the electrodes.

The RoboRoach Kit

That aside, we're happy to report that after a year of development work, we have working prototypes and we're beginning to demonstrate and distribute our remote control cockroach kits to amateurs, high schools, and universities. We're also proud to say that the project was entirely self-funded, and the total prototyping cost (not including brainpower and labor of course) over the past year was in the neighborhood of $1,000.

There are real educational benefits to these bio-robots. The same techniques used in the

⚹ BUG REPORT
A RoboRoach takes a well-deserved break after a public demo at a cafe in Woods Hole, Mass.

RoboRoach are also used in deep brain stimulation (DBS) for the treatment of Parkinson's disease and in cochlear implants to restore hearing in the deaf. These clinical devices are still relatively crude and have much room for improvement. If more and more young engineers and biologists work on the problems of neural interfaces, perhaps a renaissance in neurotechnology and treatment of nervous system afflictions can occur.

We hope that what we had to spend taxpayer money and years of our time in graduate school to study (i.e., neurophysiology of the mammalian brain) will become accessible and common high school knowledge over the next 20 years.

If you'd like to learn more, please visit us at roboroach.backyardbrains.com. ⬈

Timothy Marzullo and Gregory Gage are the co-founders of Backyard Brains, which makes neuroscience kits for students.

Timothy Marzullo (top); Emily Anthes

MENDOCINO MOTOR

See photons turned into motion with this solar-powered, magnetically levitating electric motor.

BY CHRIS CONNORS

The Mendocino Motor floats in its own magnetic field and converts light into electricity and magnetism, which are then converted into the motion of the motor. It provides the satisfaction of creating an amazing bit of technology, and the opportunity to explore magnetism, electromagnetism, electric motors, solar power generation, and personal manufacturing.

Build the base that holds the magnets and provides a bearing point for the motor. Then wind the motor coils, and solder them to the solar cells. When the motor is assembled, you'll balance it so it spins freely, and perform any troubleshooting to make it work properly.

START

1. Prepare connectors and magnets.

Download the 3D part files from makezine.com/v/31 and print the base connectors, bearing plate connectors, stator connectors, rotor block, and front and rear rotor bushings (Figure A). Clean up any rough spots with sandpaper or a utility knife.

Gather 8 ring magnets for the base, 4 for the stators, and 2 for the rotor. Use a compass to check the polarity of the magnets, and use a pencil to label them North and South.

If you have access to a laser cutter, cut the bearing plate (also at makezine.com/v/31) from ⅛" acrylic. Two designs are provided.

2. Cut 2 pencils.

Cut four 60mm segments from 2 of the pencils, using the utility knife (Figure B).

Roll the pencil beneath the knife until it cuts most of the way through, then break the pencil.

3. Build the base.

Slip 2 pencil segments into the side holes of the connectors, and 2 into the end holes, to make a frame that has 2 end connectors pointing forward, 2 pointing back, and 2 stator connectors pointing sideways (Figure C).

Slide a pair of magnets onto each end connector, with South poles facing in toward the center of the base. If you're unsure which pole

Gunther Kirsch

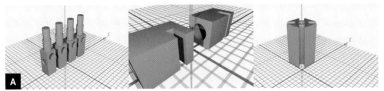

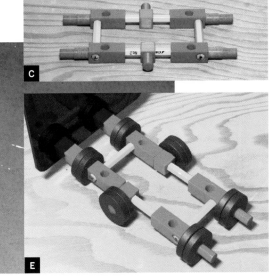

is South, just make sure the magnets at one end all face the same way magnetically, then make sure the ones on the other end also have the same pole facing inward. Remember, opposite poles attract, while like poles repel.

Slip 2 magnets onto each stator connector (Figure D), each pair pointing the same way magnetically. These will provide a stationary magnetic field around the rotor.

Slip the bearing plate connectors onto one pair of end connectors. Slide the mirror or acrylic bearing plate into the slots to complete the base (Figure E).

4. Build the rotor.

Place the pencil into the rotor block, and slide the block to about the center. Push the front bushing onto one end of the pencil. It's tapered internally, so it won't go in all the way. Then slide one magnet onto that bushing.

Push the rear bushing into a ring magnet, then slide it onto the rear of the rotor pencil. Slide the magnet further along the bushing until it "bites" down onto the pencil (Figure F).

MATERIALS AND TOOLS

3D-printed parts: base connectors (4), bearing plate connectors (2), stator connectors (2), rotor block (1), front rotor bushing with bearing point (1), rear rotor bushing (1) Download the 3D models from makezine.com/v/31.

Pencils (3) or ¼" wood dowels about 7½" long

Mirror or acrylic sheet, ⅛" or 3mm thick, about 6"×6" for the bearing plate

Ring magnets, ¼" thick × 1⅛" OD, ⅜" ID (14) RadioShack #64-1888. Three packs of 5 are needed.

Magnet wire, 30 gauge (100') RadioShack sells a 3-roll pack, #278-1345. The 30-gauge roll has enough wire for several Mendocino motors, plus you get 2 other gauges you can use for other projects.

Wooden paddle or narrow board, 1' long A paint stirrer works well.

Solar cells (4), about 2"×¾" Plastecs item #WB-18 (plastecs.com) measures 1.6"×0.8"; they're cheap but fragile, so get extras. Plastecs will make custom-sized cells, but get them with tab wires pre-soldered, because soldering them is tricky. Solarbotics #SCC2433B-MSE (solarbotics.com) measures 24mm×33mm; it's more expensive but it's encapsulated in epoxy for strength, has proper solder pads, and produces a higher voltage.

Tape, clear adhesive

Glue, all-purpose such as barge cement or fabric glue

Compass, magnetic

3D printer (optional) If you don't have access to a 3D printer, send the files to a service like Shapeways or Ponoko and they'll print and mail the parts to you.

Laser cutter (optional) to cut acrylic

Utility knife

Sandpaper, 200 grit

Soldering iron and solder, rosin core

Flux (optional) if you're soldering leads to solar cells

G

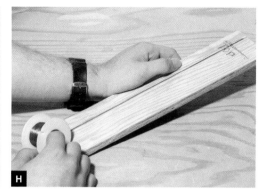

H

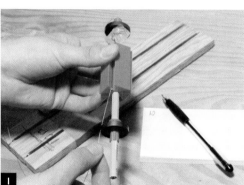

I

5. Test the magnetic levitation.

Now you'll float the rotor so you can see how the magnetic levitation works. By floating the axle and rotor block without the other parts on it, you can test and adjust without worrying about breaking the solar cells.

Test the rotor's balance by putting it into the field of the base magnets. Place the rotor so its bearing point is touching the bearing plate, and adjust the rotor magnets if necessary so that the front rotor magnet is centered *directly above the frontmost* base magnets, and the rear rotor magnet is centered *on the seams between the rear pairs* of magnets.

Now spin the rotor gently and let it turn (Figure G). It's floating! (If not, check the polarity of the magnets and try again.)

Make sure the rotor magnets don't wobble around on the pencil. If they can slip out of place, the magnetic field will force them away from their ideal position. The bushings are tapered externally, so snug the magnets up on the bushing if necessary.

If the rotor climbs up the glass, the rotor

magnets are too close to the mirror. If it jumps off the glass, they're too far away from it. Adjust them until the rotor floats evenly.

6. Measure out the magnet wire.

To measure out two 50' lengths of magnet wire, cut 2 notches in each end of the 1' wooden paddle. Wind the wire end-to-end onto the paddle, using one pair of notches to hold it (Figure H). Count 25 turns; this will measure 50'. Break the wire and secure it with tape or wrap it around the notch.

Do the same for the second 50' length. Now all your wire is measured and secured on the paddle. Loosen one end of one wire and put the paddle on the floor below you. It will unwind as you wrap the rotor.

7. Wind the motor coils on the rotor.

Wrap the first magnet wire around one end of the pencil and label this end "S1" with a piece of paper and tape, meaning "start of wire 1."

Hold the rotor in your right hand with one rotor block groove facing toward you. With your left hand, bring the wire down into the groove toward you. Wind it under, into the opposite groove on the bottom, and back over into the top groove, remaining on the left side of the pencil (Figure I). Wrap the wire around the rotor in this fashion until you've made 10 turns in the same grooves, on the left side of the pencil.

Write down the number 10. If you lose count, your notes will help you lose just a few turns instead of unwrapping the whole thing and starting over.

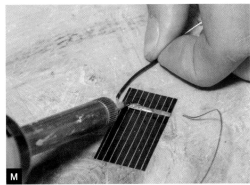

After the first set of 10, transfer the rotor to your left hand. Now make 10 turns of wire on the right side of the pencil, making sure it settles into the same grooves. When you have 10 more turns, write 20 to record this set.

Keep alternating the side every 10 turns, always staying in the same grooves. There should be 100 turns of wire on the rotor when you're done. Label the end of this wire "E1" and secure it to the end of the pencil.

Wind the second strand of magnet wire the same way, using the remaining 2 empty grooves and alternating sides every 10 turns. Start wire 2 on the opposite end of the pencil from where you started wire 1 and label it "S2." Your second wire should give the same number of turns as the first. Label its end "E2."

8. Test and trim the coil wires.

Your rotor is now wrapped with 2 lengths of magnet wire, their 4 wire ends clearly labeled S1 and E1, S2 and E2. Carefully detach the wire ends from the pencil, keeping track of which is which (Figure J).

Remove ½" of insulation from each wire end using a patch of fine-grit sandpaper. This insulation is usually colored red, but sometimes it's green or even clear. You can also scrape it off with a knife, but be careful not to break the wire too close to the rotor.

Use a multimeter set to "Continuity" to test the wires and confirm you've labeled them correctly. Then trim the wires to about 4" long and sand the ends again, so that you can solder them to the solar cells.

9. Solder leads to the solar cells (optional).

If your cells already have leads or "tab wires" soldered to them, then skip this step.

If you're using Solarbotics cells, it's easy: just solder wire leads to the positive and negative pads of each cell, the way you'd solder wire to a circuit board. You're done!

If you're using Plastecs cells with no tab wires, this step is somewhat difficult. The cells are heat-sensitive and shatter easily, so it's important to solder the joint quickly. Pre-tin the cell and the wire lead, then touch them together and briefly reheat to solder the 2. If that doesn't work well for you, try using flux.

Apply flux to the cell where you intend to solder it (Figure K), and possibly dip the end of your lead in flux as well. The positive output is the entire back side of the cell (Figure L), and the negative output is the wide stripe across the front of the cell (Figure M). Flux helps the solder to bond to the metal faster, reducing the risk of breaking or overheating the cell.

Even if you're experienced with a soldering iron, your solder joints will likely not look pretty. There may be some flux residue left, but it won't harm the cell's performance much. Just make sure the solder joint is firm.

Also, if you use solid-core wire, make sure both leads exit the cell in the same direction, running lengthwise relative to the cell. They're hard to bend later without damaging the cell.

10. Attach the solar cells.

Each solar cell has 2 leads: one coming from the front and one from the back. Arrange the

TIP: If you leave the tape in place, it will serve as a protective cover for the solar cells. Later, you can also adjust the center of balance by adding solder or another non-ferrous material between the tape and the coil.

N

O

P

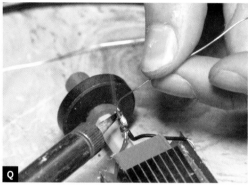

Q

cells on the rotor block so that the front lead can easily reach the back lead of the cell on the opposite side of the block.

Put a bead of adhesive on one face of the block. Barge cement, fabric glue, or other goopy all-purpose glue is ideal. You want a glue that is liquid when applied, doesn't change shape much, and is easy to spread.

Stick the first cell to the block (Figure N) and tape it in place with 100mm of clear adhesive tape, starting on the bare face of the block adjacent to the cell, wrapping over the cell, attaching to the next bare face.

Glue the second cell to the face opposite the first cell, ensuring that its positive (back) lead is near the negative (front) lead of the first cell, and vice versa.

Bring the rest of the tape over the second cell, to hold both cells until the glue dries.

Repeat this process to attach the remaining 2 solar cells (Figure O), always making sure that the tabs from adjacent cells do not touch each other.

11. Connect the solar cells and coils.

Bend all the leads so that each negative (front) lead is mechanically connected to the positive (back) lead of the cell *on the opposite side* of the block (Figure P).

Solder all 4 connections, making sure you're connecting each pair of cells on opposite sides of the block, not cells that are next to each other (Figure Q).

Connect each of the coil windings into the circuit as shown in Figure R. Each coil is driven by its own pair of solar cells. Connect the first coil to one pair of cells by soldering the coil leads into the (+/−) connections you just made between those cells. Repeat with the second coil and second pair of cells.

12. Test the motor in the light.

Place your fully assembled and balanced motor in bright light to see it work (Figure S). The motor turns best in sunlight. Halogen and incandescent lights also work very well. Fluorescent light usually doesn't work well.

Your motor should self-start if it's balanced

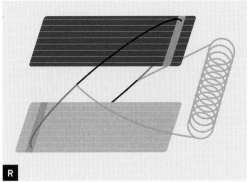

properly and the magnets are placed correctly with no wobble. Sometimes you can coax a motor into running by moving the light rhythmically and getting it to rotate. You can also gently spin the motor to get it started. Be careful not to spin it too fast, or it may jump out of the magnetic field and break.

The stator magnets provide a stationary force that the motor turns against. Your motor may turn without them, but the fastest-running motors I've seen used stators.

13. Troubleshoot the motor.

Balance is a common problem. Remove the stator magnets, then turn the rotor ½ turn and let go (don't spin it). If it rotates back to its original position instead of rotating forward, you can add weight to the highest part of the circle. Try bits of brass or solder. Don't add nails or other steel weights, because they're made with iron, attractive to magnets.

Magnets must be accurately placed. If they can wobble out of position, they will. If needed, neatly tape the base magnets to keep them aligned straight up and down. Make sure all South poles are facing inward. If needed, add tape to the bushings to keep the rotor magnets from wobbling. Again, the front rotor magnet is centered over the frontmost base magnet, and the rear rotor magnet is centered over the rear *pair* of base magnets.

Solar cells — Make sure you have continuity between each pair of opposite cells, and not between adjacent cells. If your continuity is off, you may have to rewire the connections. If a cell is connected to an adjacent cell, then reconnect it to the opposite cell.

Reversed polarity — If everything above is correct, and the rotor still just wobbles back and forth in the light, try this technique. Neatly cut one pair of coil wires from one pair of cells. Re-sand the ends of the wires, then solder them onto the cell leads opposite from the leads they were on originally. This changes the direction of the electricity flowing through that particular coil. If your problem was reversed polarity, this should fix it.

14. Show off your Mendocino Motor!

When I show the Mendocino Motor to electrical engineers or physics teachers, I have fun hearing them think out loud about how it works.

Place your Mendocino Motor in a sunny window, and you'll see it start up when the sunlight is bright enough, and slow to a stop when the day fades. Put it on your desk and watch it turn in the light of your lamp as you work. Try different types of bulbs for differing results in speed.

The Mendocino Motor has little torque or turning power, so it's a challenge to harness its motion for other uses. You could fashion a fan on the rotor axle to move some air. (Stick more magnets onto the stators to increase torque.) What other uses can you think of? ⬛

MAKE contributor Chris Connors is a teacher who loves to learn with curious people who are interested in inventing the future. His students have made hundreds of Mendocino Motors over the years.

26 HACKS IN 48 HOURS

The creative explosion of Science Hack Day.

BY ARIEL WALDMAN

"Overwhelming." "Unexpected." "Joyful." "Serendipitous." "Inspiring." "Fun." These are typical words used to describe the experience of participating in a Science Hack Day, a 48-hour all-night event where anyone excited about making weird, silly, or serious things with science gathers to see what they can prototype within 24 consecutive hours. Designers, developers, scientists, and all science enthusiasts are welcome to attend — no experience in science or hacking is necessary, just an insatiable curiosity.

In November, San Francisco hosted 150 science hackers for a weekend of making — ten of whom were interested in creating a similar event in their home cities of Berlin, Cape Town, Chicago, Dublin, Mexico City, Nairobi, Reykjavik, Sao Paulo, Tokyo, and Vancouver. Some arrived with ideas of what they might want to do, while others showed up with nothing more than enthusiasm. People organically formed multidisciplinary teams over the course of the weekend: particle physicists teamed up with designers, marketers joined forces with open source rocket scientists, writers collaborated with molecular biologists, and developers partnered with schoolkids.

The result of this weekend of intense hacking and collaboration was 26 remarkable hacks, spanning across all types of science, including oceanography, neuroscience, space exploration, seismology, biotech, and particle physics, among others.

Science Hack Day is about mashing up ideas, mediums, industries, and people to create sparks for future ideas and collaborations. Here are some of the hacks created at the most recent and previous Science Hack Days.

Syneseizure

Wouldn't it be cool if you could feel sight? That's what one team of science hackers sought to explore, creating a mask that simulates synesthesia, a condition where senses get mixed up (e.g. associating colors with numbers or seeing ripples in your vision resulting from loud sounds). The team wanted to simulate a synesthetic sensation by mashing up sight (via a webcam) with touch (via vibrating speakers).

Syneseizure is a fairly creepy-looking hack. They attached 12 vibrating speakers inside a head mask sewn from a pattern they found online and wired them to an Arduino and a webcam. The result is an all-encompassing head mask that vibrates on different areas of your face, corresponding with different visual information picked up by the webcam, thus creating the feeling that areas of a room are lighter or darker as you navigate around. Learn more about the making of Syneseizure at syneseizure.wordpress.com.

Particle Windchime

Instead of seeing visualizations of subatomic particle collisions, what if you could hear them? Matt Bellis, a Stanford particle physicist, teamed up with a focused group of hackers and experimented with mapping

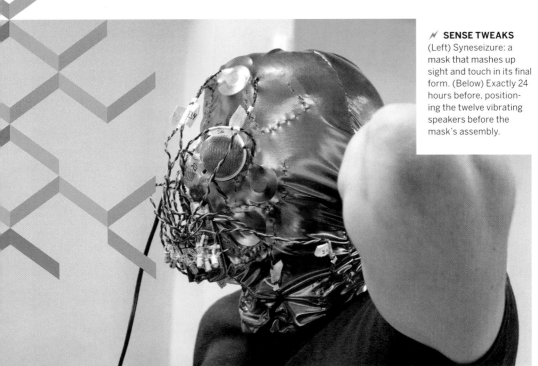

Matt Biddulph (top); Gregory Hayes (bottom)

SENSE TWEAKS
(Left) Syneseizure: a mask that mashes up sight and touch in its final form. (Below) Exactly 24 hours before, positioning the twelve vibrating speakers before the mask's assembly.

measured small lines, and as a result of being exposed to this fun hack, Bellis wrote a proposal for how to detect cosmic rays in a cloud chamber using similar techniques. Download the Particle Windchime to run on your computer at mattbellis.com/windchime.

DNAquiri

What does DNA taste like? Aside from the fact that DNA is very small, the materials needed to extract it often aren't edible, or if they are, they're not as delightful as a cocktail.

Despite the copious amount of food present at Science Hack Day, a band of biohackers were hungry for more. They sought to craft a recipe for extracting strawberry DNA that didn't require indigestible ingredients and could also double as a cocktail.

Using strawberry puree and some very strong alcohol, the biohackers were able to extract the strawberry DNA into polymer clumps you could see with the naked eye. The final cocktail was definitely something that could knock you off your feet, but it has paved the way for more delightful science-

particle collision data with a variety of sounds. The result is an amusingly awkward symphony of science that you can control via a web interface. While on the surface it is a "cute" hack, Bellis thinks it could be considered as a sort of augmented diagnostic tool for accelerator laboratories to use for detecting problems in the accelerator.

Subsequent hacks at Science Hack Days have also gone on to inspire particle physics in unexpected ways. One team created a beard-detecting hack using a USB microscope paired with a computer vision library that

based delicacies. The recipe is available at makezine.com/go/dnaquiri.

✎ DNAquiri: a cocktail of extracted DNA. (Opposite) ISS Globe: a globe displaying the real-time location of the International Space Station, and a makeshift wind tunnel used to create the Isodrag Typeface.

Quake Canary

What if our phones could broadcast "earthquake!" faster than we could tweet it? We are all cyborgs, after all, carrying devices that extend us physically and mentally. Those devices are more than just phones, though, they're also sensors — ubiquitous, cheap, widely-distributed sensors that many scientists are eager to tap into.

The Quake Canary hack put this concept to a test: can our smartphones detect earthquakes accurately? It seems so. By prototyping a proof-of-concept, the team demoed the ability for networked phones to detect earthquakes and instantly send data to the U.S. Geological Survey — potentially giving areas in danger early warning signals quicker.

The hack has now blossomed into a project that's teaming up with machine learning and time-domain informatics experts from the University of California, Berkeley. Prototypes with improved algorithms have been created

by the team in the last few months, and they are preparing for expanded deployment of the project along California's Hayward Fault.

ISS Globe

Can you check up on space travel while you're relaxing at home or busy at work? Astronauts are continuously orbiting the Earth — sometimes you can see a fair glint of their spacecraft, the International Space Station (ISS), overhead on a clear night when they happen to fly past your location. What if you could always see where they are without going outside or opening your laptop?

Inspired by previous Science Hack Day ideas like the Near Earth Asteroid Lamp, a lamp that would light up each time an asteroid passed by the Earth, the ISS Globe continuously shows you where the ISS is overhead. The team of science hackers used a translucent globe, two hobby servos,

Matt Biddulph

> **Science Hack Day is about mashing up ideas, mediums, industries, and people to create sparks for future ideas and collaborations.**

a MakerBot for 3D printing a few gears, a Teensy microcontroller, and a laser that was mounted inside the globe. Firmware controls interaction between a laptop and the servos, and a Python client controls interaction between the microcontroller and the servos.

The end product is a globe with a glowing red dot that shows you where the ISS currently is throughout the day. Learn more about the making of the ISS Globe and find the source code and API at makezine.com/go/issglobe.

Isodrag Typeface

Typefaces often strive for visual consistency in their design, but what if they took their visual cues from the physical world?

The Isodrag Typeface is a font designed by aerodynamics. This hack used a makeshift wind tunnel and recorded the aerodynamic drag of each uppercase sans-serif letter. The weight of each letter was altered until all letters were recorded as having equal drag. For example, the letter "I" has low aerodynamic drag, so it was modified to be very thick. The

product is a typeface that is aerodynamically consistent. You can see the full typeface set at twitpic.com/7dpbrd.

These hacks are very much in the spirit of Science Hack Day's mission to get excited and make things with science. By creating new ways of interacting with, contributing to, and deconstructing science, the amazing things that emerge often influence and impact the scientific field in unexpected yet delightful ways.

Science Hack Days are being planned in dozens of cities all around the world, and if there's not already one coming to your community, you're encouraged to create one! Instructions for how to create a Science Hack Day in your city and a list of upcoming events are at sciencehackday.com. ◪

Ariel Waldman is a global instigator of Science Hack Day and the founder of Spacehack.org, a directory of ways to participate in space exploration. Her website is at arielwaldman.com.

RENEGADE RESEARCH

♥ BUILDING BLOCKS OF LIFE

IS THERE NO LIMIT TO THE VERSATILITY OF LEGO? Researchers at the University of Cambridge are using Lego Mindstorms sets to automate a process for creating synthetic bone tissue. Ph.D. students in the Department of Engineering built robotic cranes out of Lego pieces and motors that repeatedly dip tissue compounds in solutions of calcium/protein and phosphate/protein to build up the artificial bone samples. Lego was also simpler and cheaper than other off-the-shelf options.
makezine.com/go/legobone

—*Craig Couden*

Khaow Tonsomboon

CHECK YOUR HEAD 💀

Forensic scientists and investigators have a new software tool to help unidentified human remains speak for the dead. 3D-ID goes beyond previous facial reconstruction methods by identifying both the sex and ancestral background from measurements of a skull. After digitally mapping 34 coordinate points on the skull, the geometric data is compared to a library of 1,300 individuals representative of a wide range of populations. The software, written in Java, is available to download for free. www.3d-id.org   —CC

BIOPUNKS UNITE!

DIYbio is a global network that seeks to connect regional scientists with citizen science groups to promote biohacking for all, with an emphasis on the exchange of knowledge. With the power of open source and open access, these "biopunks" believe they can speed the process of biotechnology innovation. The community actively discusses the ethics of biotechnology and promotes tenets like peace, community, education, modesty, safety, and transparency. To join a discussion, read biohacking news, or find a bio-hackerspace near you, visit diybio.org.   —*Laura Cochrane*

HACK TO THE FUTURE

Traditional lab equipment for spatial frequency domain imaging (SFDI), an optical diagnostic technique, is staggeringly expensive, but students at the Beckman Laser Institute in Irvine, Calif., are putting their DIY skills to work. One example, grad student Alexander Lin has hacked a $300 Aaxa LCD projector to replace the traditional development kit that costs upward of $6,000. He's using this technology to help understand what happens to the brain in Alzheimer's disease. www.bli.uci.edu
          —*Goli Mohammadi*

North Carolina State University (3D-ID); Mackenzie Cowell (DIYbio); Alexander Lin (LCD hack)

APPETITE FOR INFORMATION

Wu Heng, a 25-year-old graduate student in Shanghai, began tracking food safety scandals in China after he discovered that his favorite restaurant had been serving "fake beef" (pork doused in additives to alter its texture and color). Stomach-churning news stories like one about cooking oil extracted from septic tanks compelled him to act.

Heng and 33 volunteers combed through thousands of reports, listing them according to date, area, and food category. In June 2011, they launched a website to document and map incidents nationwide. Though collected from secondary sources, it's the first effort of its kind to present Chinese citizens with a comprehensive overview of the problem. zccw.info
—Wendy Becktold

> # " Danger warnings abound, but that means you're doing something right.

QUANTUM PHYSICS FAM

They say the family that performs DIY quantum physics experiments together, writes a book about said experiments together. Such is the case with father/daughter team David and Shanni Prutchi of diyphysics.com. Projects in the book range from depicting light as both particle and wave to demonstrating quantum entanglement. The pair use their website to share projects they've built, including a Cockcroft-Walton generator that kicks out high DC voltage from low-voltage AC. Danger warnings abound, but that means you're doing something right.
—CC

2004

2005

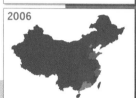

2006

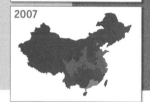

2007

2008

2009

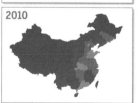

2010

2011

Wu Heng (zccw.info); David Prutchi (quantum physics)

:) :(

GEOMAGNETIC GENIUS

What could be cooler than monitoring magnetic storms from your own backyard? Building the detection device yourself and housing it in PVC. Alex Avtanski's homemade magnetometer translates the movement of a laser beam into recorded fluctuations of Earth's magnetic field, which are caused by the sun. See his complete how-to and check out real-time graphs of fluctuations at avtanski. net/projects/magnetometer.

—*Paul Mundell*

TRY THIS AT HOME

The *Illustrated Guide to Home Biology Experiments* (O'Reilly Media) had me at the tagline, "All Lab, No Lecture." Not a textbook (though it links to freely down-loadable texts), the collection of over 30 experiments is great for getting hands-on with science on your own terms. I might not do all the experiments myself, but I'm recommending this to all my homeschooling parent friends. I'll gladly drop by to help and play. makezine.com/go/homebio

—*Gregory Hayes*

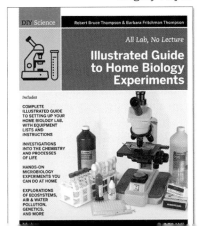

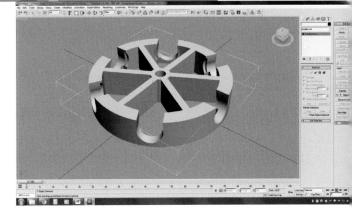

DREMELFUGE

Billed by its creator as the world's cheapest super-centrifuge and only open source hardware centrifuge, the Dremelfuge puts lab-strength high-g in the hands of the DIY bio masses. Combined with a Dremel 300's maximum speed of 33krpm, the Dremelfuge can spin hapless experiments at over 50,000g, theoretically enough to achieve cell frac-tionation. The 3D-printable files are available for free on Thingiverse, or Shapeways sells pre-printed models for under 50 bucks; both come with my favorite warning, to use at your own risk. indie biotech.com/?page_id=16

—*GH*

HOT IDEA

⚡ BIG PROBLEM, NANO-SOLUTION

Forget annihilating the rival football team — Cupertino, Calif., high schooler Angela Zhang designed a nanoparticle that kills a much more formidable opponent: cancer cells. Her grandfather and great-grandfather died from the disease, which motivated her to search for a cure. Zhang's technique involves mixing a polymer with cancer medicine and attaching it to nanoparticles, which then attach to cancer cells. The nanoparticles are identified by MRI, and doctors aim an infrared light at the tumors, melting the polymers to release the medicine, which kills cancer cells and leaves healthy cells unharmed. In tests, mouse tumors almost completely disappeared. makezine.com/go/angelazhang   —LC

Want to conserve energy?

A thermal-imaging camera will help you detect heat leaks in your wall. Problem is, they cost thousands of dollars. Now the Public Laboratory for Open Technology and Science (PLOTS) has developed a DIY thermal flashlight that does the same thing. It consists of a non-contact infrared thermometer that reads the temperature of an object and an RGB LED that "paints" the surface in red light if hot (75°F or more) and blue if cold (60°F or less). Housed in a VHS tape case, the entire device costs less than $100 (though you also need a camera that can take a long-exposure photograph to record the images). Find instructions at publiclaboratory.org/tool/thermal-photography.   —WB

DATA OVERFLOW

Using only an Arduino Uno and a few extra components, a group of dedicated conservationists have devised a system to monitor the water in New York City's Gowanus Canal. Based in Brooklyn, the group Don't Flush Me set up their gadget to test water temperature and conductivity, two variables that change drastically immediately after a sewer overflow. With an attached shield, the Arduino broadcasts the data so it can be interpreted from any computer. makezine.com/go/gowanus   —PM

LAB AT HOME

TABLETOP BIOSPHERE

Create a sealed ecosystem that supplies a freshwater shrimp "econaut" with food, oxygen, and waste processing for a desktop journey of 3 months.
makeprojects.com/project/t/329

YAGI ANTENNA

Tune in to space with a handheld multiband radio while tracing the orbit of a satellite or the ISS with a homemade yagi antenna.
makeprojects.com/project/h/623

COUNTRY SCIENTIST

In each issue of MAKE, renowned scientist and author Forrest M. Mims III teaches the ways of citizen science, from data mining to image analysis.
makezine.com/countrysci

GEIGER COUNTER

Build a radiation detector that clicks, flashes, logs radioactivity levels, and shares its data with the world.
makeprojects.com/project/g/1821

MYCOLOGY LAB

Use an off-the-shelf home air purifier to make your own laminar flow hood to culture and grow mushrooms.
makeprojects.com/project/h/242

HIGH-RES SPECTROGRAPH

Get inexpensive, lab-worthy spectrum analysis using a digital camera and plumbing parts.
makeprojects.com/project/h/658

IMAGING "SATELLITE"

Snap aerial photos from 300' up by suspending a hacked drugstore camera from three helium balloons.
makeprojects.com/project/b/307

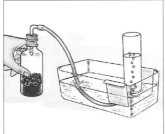

PNEUMATIC TROUGH

Collect pure gas samples with this handy, homemade piece of classic chemistry lab equipment.
makeprojects.com/project/p/431

VAN DE GRAAFF GENERATOR

Demonstrate the triboelectric effect and shoot electrical sparks with a soda can, rubber band, and PVC pipe.
makeprojects.com/project/s/2072

FOR MORE GREAT PROJECTS LIKE THESE, GO TO MAKEPROJECTS.COM/C/SCIENCE.

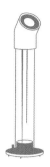

SOUND-O-LIGHT SPEAKERS

Surprisingly simple PVC pipe speakers are clear shining performers.

By William Gurstelle

While tidying up my workshop, I found some clear 3" PVC pipe left over from a spud gun project (the Nightlighter 36 taser-powered potato cannon in MAKE Volume 04). Clear PVC is one of my favorite building materials, but it's expensive, and since the last thing I need is another potato cannon, I wanted to come up with a project that would make good use of its unique qualities.

Clear PVC is stiff and dense, which makes it excellent material for audio speaker cabinets. I had seen uniquely shaped speakers made from regular white PVC, so I wondered if clear tubing could make decent-sounding cabinets that also generate lighting effects.

I connected an iPod (playing ZZ Top's "La Grange") to a 20-watt amplifier and a small speaker, and played around with different configurations of LEDs on the speaker wire. The best visuals, I found, came from simply connecting 3 ultra-bright LEDs in series, in parallel with each speaker. Voilà! At moderate volume and above, the same audio signal both drove the speaker and pulsed the LEDs in time with the music — and I discerned no difference in the speakers' sound with or without the LED load.

Introducing the Sound-O-Light Speakers. They're easy to build, they get surprisingly good sound out of their single 3" drivers, and they look hella cool.

William Gurstelle is a contributing editor for MAKE magazine. Visit williamgurstelle.com for more information on this and other maker-friendly projects.

SET UP: p.83   **MAKE IT: p.84**   **USE IT: p.87**

GET THE LEDS OUT

Audio speaker design is an exacting science with a bit of art mixed in, and the mathematics around driver and cabinet characteristics and behavior gets complicated fast. The Sound-O-Light speaker bypasses all of this with a very basic design, and its LEDs simply connect alongside the driver.

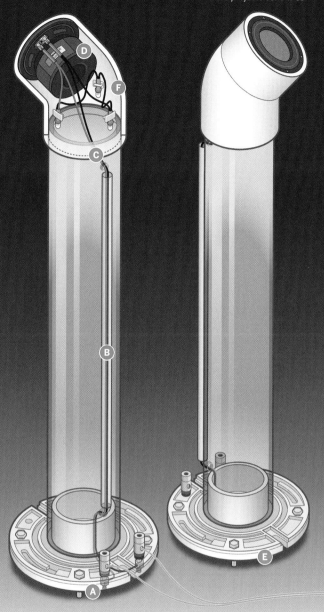

(A) An audio source and amplifier send audio signals via speaker cable to audio binding posts at the base of each speaker.

(B) An aluminum tube running up the back of the speaker cabinet carries the speaker cable up to the driver and LEDs inside the top. This keeps the inside of the speaker cabinet empty and free of wires, for visual and acoustic neatness.

(C) The speaker cable splits at the top to power both the speaker driver and LEDs.

(D) The speaker driver inside the 45° pipe elbow produces sound by vibrating air in front of its cone, while also vibrating the air inside the cabinet from its back, 180° out of phase.

(E) Depending on whether the speaker has a *bass reflex* or *acoustic suspension* design (see below), sound inside the cabinet may also exit at the speaker's PVC flange base.

(F) Three LEDs inside the 45° pipe elbow shine light down the clear PVC tube cabinet. Powered by the audio signal, they blink and glow following the frequency and volume of the music.
At low volumes, the raw audio signal isn't enough to make the LEDs glow, so an optional 9V battery and transistor circuit inside the speaker (as shown on page 85) gives the LEDs more juice to make them more active.

ACOUSTIC SUSPENSION

With **acoustic suspension** speaker designs, the cabinet is sealed. This makes the air inside act like a spring that pushes or pulls the driver cone against its direction of travel and toward its midpoint, as it tends to equalize with outside air pressure. This design doesn't generate the greatest volume but can enhance a speaker's accuracy.
To make an acoustic suspension cabinet, seal the bottom opening of the pipe flange with a plastic disc.

BASS REFLEX

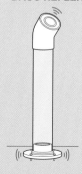

Bass reflex speakers, in contrast, have vents that redirect the sound inside the cabinet back outside, reinforcing the sound produced by the front of the driver. This can increase volume, especially for the low notes that the back of a driver cone generates more effectively.
For a bass reflex cabinet (shown above), leave the bottom open using the bolts as spacers between the flange and the floor. Try it both ways to determine which you prefer.

Timmy Kucynda

SET UP.

A

B

C

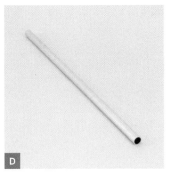

D

E

MATERIALS

A. Speaker drivers, 3" diameter, round (2) I used HiVi B3N drivers, which are popular, cost about $10 each, and use standard car-audio spade connectors. If you have a different 3" driver in mind, by all means give it a try.

B. PVC pipe, clear, 3"×20" lengths (2) Some sources sell this by the foot while others only sell 10' lengths. Since this pipe costs roughly $15/foot, it makes sense to find a supplier that can sell the exact 40" quantity required. Check local industrial plastics suppliers first, but there are also several online sources.
» **PVC pipe flange fittings, 3", white (2)** ABS fittings may also work, but ABS is not as stiff or dense as PVC, which may affect audio quality.
» **PVC pipe elbow fitting, 45°, 3", white**

C. Audio binding posts (4) typically sold in pairs

D. Aluminum tubing, ½" outside diameter (OD) × 17½" long (2)

E. LEDs, 5mm, ultra-bright (6) your choice of color(s); they can be different for each speaker.

» **Spade connectors, female, .110" (4)** standard car audio connector
» **Cable clips, adhesive-backed, plastic, sized to hold coaxial cable (6)** for holding LEDs. You can also use hot glue.
» **Speaker cable, 2-conductor, 6'** or other paired wire. We used a red/black pair to help identify the polarity for the LED circuit.
» **Hookup wire, insulated, 18–22 gauge, 4'**
» **Silicone adhesive** in squeeze tube

For bass reflex speakers:
» **Bolts, #12 round head, ¾" long, with matching nuts (8)**
» **PlastiDip or 6" hardwood square (optional)** to prevent bolts from scratching floors or furniture

For acoustic suspension speakers:
» **PVC or other plastic sheet, ¼" thick, 1' square minimum** Look in the window and door aisle at your home improvement store.
» **PVC cement**
» **Silicone caulk**

F. For powered/transistorized option:
» **Snaps for 9V battery (2)**
» **9V batteries (2)**

F

» **Transistors, NPN, TIP31 type (2)**
» **Resistors, 220Ω (2)**
» **Switch, on-off toggle**
» **Perf board**

TOOLS

» **Saw for cutting PVC**
» **Soldering equipment**
» **Drill with ¼" bit**
» **Cresent wrench, adjustable**
» **Needlenose pliers**
» **Wire strippers and cutters**

Gunther Kirsch

MAKE IT.

BUILD YOUR SPEAKERS

Time: An Afternoon
Complexity: Easy

1. CUT AND GLUE THE TUBES

1a. Cut the clear PVC and aluminum tubing to length. Drill two ¼" holes in each PVC tube, 1" from each end and in line with one another, parallel to the tube.

1a

1b. Use a thin bead of silicone adhesive to attach the aluminum tube to the PVC pipe between the 2 holes. Let dry completely.

2. CONNECT THE WIRING

2a. Cut and strip the ends of four 6" lengths of hookup wire (8 lengths total for both speakers). Solder 3 LEDs together in series, with wire leads in between and at each end. Orient all LEDs the same way, with neighboring LEDs connected anode-to-cathode (longer leg connected to shorter leg).

2b. Cut a 32" length of speaker cable or other paired wire, peel the wires apart at each end, and strip the ends. Crimp and solder a female spade connector to each wire at one end of the cable.

NOTE: These instructions describe the construction of a single speaker. Perform each step twice to make a pair.

1b

⚠ TIP: An LED's cathode is indicated by both a shorter leg and a small flat on the side of its plastic body.

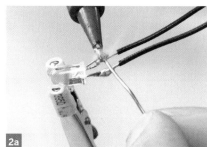

2a

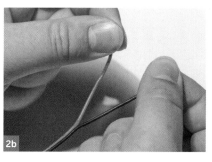

2b

2b

2c. (Optional) If you want to power the LEDs with 9V batteries so they flash at lower volumes, wire the circuit shown here.

2c

2c

2d. Wrap each stripped wire end of an LED chain around the copper base of each spade connector, and solder in place. (With the battery circuit, connect the LED's cathode end and the transistor's collector, the middle leg.) The LED anodes and cathodes can be oriented in either direction on the speaker wire.

2e. Plug the female spade connectors onto the speaker terminals and solder in place.

2f. Test the wiring by connecting the speaker/LED assembly to an amplifier and playing music. During loud passages (or more often, if powered), the LEDs should pulse in time to the music. If the LEDs do not light, check the quality of the soldered connections and make sure all LEDs are connected positive to negative.

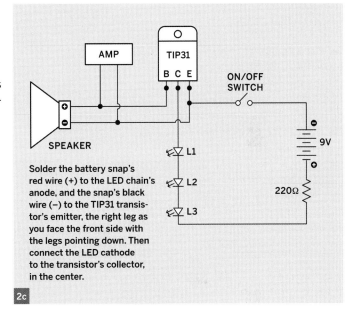

Solder the battery snap's red wire (+) to the LED chain's anode, and the snap's black wire (−) to the TIP31 transistor's emitter, the right leg as you face the front side with the legs pointing down. Then connect the LED cathode to the transistor's collector, in the center.

2c

3. FINAL ASSEMBLY

3a. Attach the cable clips to the interior of the PVC elbow fitting so they will hold and point the LEDs down the PVC pipe, spaced 120° apart. They should sit well up inside the fitting to avoid being loosened when you insert the pipe. You can also secure them with hot glue (Step 3c).

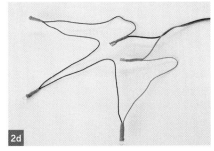

2d

2e

2f

3b. Place the speaker into the elbow, center it, and then secure it in place with silicone sealant around the perimeter. Let the sealant dry completely.

3c. Turn the elbow upside down and clip or glue the LEDs inside, making sure they will point down the tube when the speaker is on top.

3d. Prepare the PVC flange to be the base of the speaker.

For a *bass reflex cabinet* (shown here), run 4 bolts down through the mounting slots around the flange, securing them with nuts on either side.

If your flange lacks slots, drill ¼" holes. For an *acoustic suspension cabinet*, use a saber saw or similar tool to cut two 5" rounds out of ¼"- thick PVC. Glue them into the underside of each flange, then seal gaps with silicone caulk.

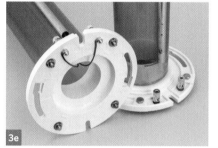

3e. Drill two ¼" holes in the flange and install audio binding posts with supplied nuts.

Thread the speaker wire dangling from the PVC elbow out through the top ¼" hole in the clear PVC pipe, down through the aluminum tube, and back in through the bottom hole in the PVC pipe.

3f. Press the elbow firmly onto the top of the clear pipe, and press the flange onto the bottom. Temporarily rest the speaker on its side, wrap a speaker cable wire around each binding post terminal, solder in place. You're done!

TEST BUILDER:
Daniel Spangler,
MAKE Labs

USE IT.

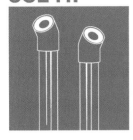

MUSIC SERVED PIPING HOT

Filling the Pipe

At low volumes, the Sound-O-Light LEDs won't glow, but at moderate volumes and above, they will pulse with the sound — the louder, the brighter. As with any speaker pair based on small drivers, you can fill out the sound by adding a subwoofer, and please let me know if you build one that glows.

More Plastic Pipe Speakers

Eric Nelson, Rob Sampson, and others make amazing-looking speakers from plastic pipe, following more advanced designs than the Sound-O-Light. Nelson uses PVC to make desktop-sized acoustic creations (Red Lobsters, on the right), while Sampson makes larger speakers using standard black ABS plastic (Drooping Didgeridoo, left) — including "quarter wave" chambers that resonate well because their unfolded length exceeds ¼ of the wavelength in air of the lowest frequency they carry, which is about 9½' for a 30Hz rumbling bass tone. ◪

➕ Visit makeprojects.com/v/31 for a video overview of the Sound-O-Light Speakers that shows them in action.

Rob Sampson (Drooping Didgeridoo); Eric Nelson, ikyaudio.com (Red Lobsters)

ROCKET GLIDER
The classic toy, remade.
This folding wing glider rockets up ...
then glides back down!

By Rick Schertle

When I was a kid, I remember my dad talking about a seemingly magical balsa-wood "rocket glider." With the wings folded back, the glider shot into the air from a handheld rubber-band catapult. As it soared upward, wind resistance held the wings back, and when the glider reached its peak, the wings popped open for a long and graceful glide down.

When the glider began to wear out — this part was especially exciting to me — he would attach a firecracker to it and launch it into a shower of balsa-wood confetti glory.

Ten years ago I found a company that sold these "rocket gliders" for ten bucks, and I bought a few. Wow, they were super fun! I was used to hand-tossed gliders that flew 20 or 30 feet. These new ones zipped straight into the sky 60 or 70 feet! My brother-in-law Brendan and I played with them like kids for hours.

Recently I discovered one in my basement. Online, I learned that the glider was out of production, the patent long expired. So, using my one remaining glider, with help from TechShop, I began reverse engineering the project and now present it here as a how-to.

Along with MAKE, I've also developed it into a handy kit that's ready to fly in about an hour. Like so many fans of this toy over the years, you can now pull the glider back on the catapult, aim straight up, and let 'er rip!

MAKE contributing writer Rick Schertle (schertle@yahoo.com) teaches middle school in San Jose, Calif., and designed the Compressed Air Rocket for MAKE Volume 15. With his wife and kids, he loves all things that fly.

| SET UP: p.91 | MAKE IT: p.92 | USE IT: p.97 |

Gregory Hayes

HISTORY REPEATS Schertle and his son launch their Rocket Glider high in the summer sky.

ROCKETS UP, GLIDES DOWN

With wings (A) folded back, the glider is launched from a handheld catapult (B). Wind resistance holds the wings back until, at apogee, they flip open for a gentle glide back down to the ground.

The magic lies in the wing pivot (C). This clever piece lets the wings tilt forward to vertical, then fold back against the fuselage (D). Tension from a rubber band (E) unfolds the wings forward and pivots them back to the horizontal position.

The straddle wire (F) holds the 2 wings together using the aluminum wing clips (G). A penny (H) slips into a slot in the nose to add weight for correct balance.

1. Launch. With its wings folded in the vertical plane, the glider is launched straight upward like a dart.

2. Open. Near maximum height, the glider stalls and the wings open, still in the vertical plane. This brakes the aircraft to begin glide phase.

3. Glide. The wings pivot into the horizontal plane, and the glider sails on.

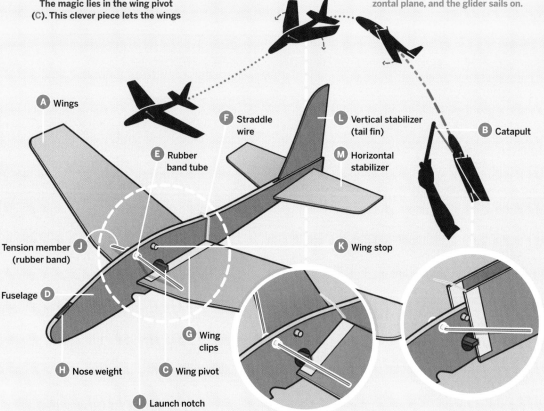

- **A** Wings
- **F** Straddle wire
- **L** Vertical stabilizer (tail fin)
- **B** Catapult
- **E** Rubber band tube
- **M** Horizontal stabilizer
- **J** Tension member (rubber band)
- **K** Wing stop
- **D** Fuselage
- **G** Wing clips
- **H** Nose weight
- **C** Wing pivot
- **I** Launch notch

GLIDER GLORY

Jim Walker was an innovator of control line airplanes, a sonic control glider, and even a remote control lawn mower. He developed and patented his original toy folding-wing glider, the Army Interceptor, in 1939, and began producing them in Portland, Ore. During the 1940s they were sold in toy and dime stores nationwide.

Scott Griffith, historian for Walker's collection, shared more of the story.

During World War II, Walker's gliders caught the interest of the U.S. Army. He developed a "Military Launcher" to catapult the gliders nearly 300' in the air as target planes for artillery practice (look for a how-to in MAKE Volume 32). The company worked 24 hours a day producing these during the war. An astounding 232 million Army Interceptors were produced during the lifetime of the American Junior Aircraft Company.

Walker's friend Frank Macy often watched him fly his models in Portland parks. Macy was so inspired that he became the preeminent American Junior historian, and continued to produce the Army Interceptor in small quantities until his death in 2009.

In 2004, R/C hobbyist Paul Bradley added micro R/C units to the newer "404 Interceptor." He believes much lighter-weight models are now possible with current micro R/C tech.

SET UP.

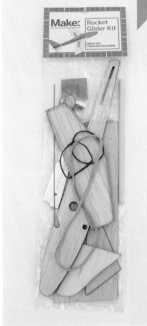

A

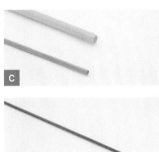
C

F

B

E

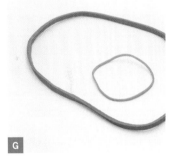
G

MATERIALS

The MAKE Rocket Glider Kit, item #MKRS2 from Maker Shed (maker shed.com), **includes all of the following glider parts, pre-cut.**

You can also buy the following materials separately and cut the parts yourself, either with a laser cutter using our DWG or CDR files, or by hand using our printable-to-scale PDF files, all of them downloadable for free at makeprojects. com/v/31.

A. Balsa wood sheets: 1/16"×6"×12" (1), and 3/16"×3"×12" (1) for the wings and the fuselage, respectively

B. Hardboard or acrylic sheet, 1/8" **thick,** scrap for the wing pivot. This is a very small piece about 1"×1/2".

C. Wood dowels: 1/8" diameter, 1/2" long (1), and 1/4" diameter, 6" long (1)

D. Stiff wire, 61/2" long I used lightweight 18-gauge floral stem wire. A straightened large paper clip is the right gauge but might be a bit too short.

E. Aluminum sheet, 0.012" thick, 21/2"×13/4" Get it from a hobby store, or cut apart a soda can.

F. Plastic tubing, rigid, 3/16" (4.8mm) OD, 1/2" length I used part #226 from evergreenscalemodels.com; you can get creative with something similar.

G. Rubber bands: #16 (1) and 7" (1) for the wing pivot and the catapult, respectively

You'll also need:
» Penny coin
» PlastiDip synthetic rubber coating (optional)
» Staples (2) standard desk variety
» Glue, white
» Glue, polyurethane such as Gorilla Glue

TOOLS

» **Laser cutter (optional)** if you're laser cutting
» **Hobby knife, and scroll saw or coping saw** if you're hand cutting
» **Tinsnips or strong scissors** if you use a soda can
» **Drill and drill bits:** 3/8", 3/16", 1/8", and 3/64"
» **Razor blade or box cutter**
» **Wire cutters**
» **Sandpaper, very fine grit**
» **Vise (optional)** but helpful
» **High-speed rotary tool with cutting wheel** e.g., Dremel
» **Needlenose pliers**
» **Metal file, small** for deburring

Gunther Kirsch

MAKE IT.

BUILD YOUR ROCKET GLIDER

Time: A Couple Hours
Complexity: Easy

1. CUT THE PARTS

If you have the kit, you can skip this step and any other cutting instructions.

1a. Use ³⁄₁₆" balsa for the fuselage and ¹⁄₁₆" balsa for the wings. For best results, cut the horizontal stabilizer with the grain going lengthwise, and cut the vertical stabilizer (tail fin) with the grain going from top to bottom.

For hand cutting, the ³⁄₁₆" balsa cuts well on a power scroll saw with a new blade. The bottom splinters a little, but can be sanded with fine sandpaper. I don't recommend cutting the ¹⁄₁₆" balsa with the scroll saw; instead, use a very sharp hobby knife, cut slowly with multiple passes, and use a metal straightedge when possible.

1b. Cut the wing pivot from ⅛" acrylic or hardboard, by hand or laser, using the corresponding downloaded file. Drill tiny holes with a ³⁄₆₄" drill bit in the pivot in the exact location shown on the template.

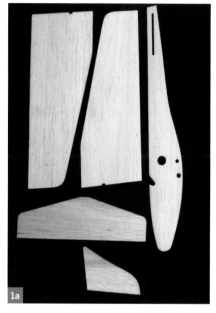

1a

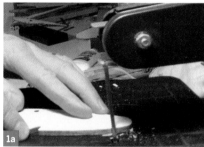

1a

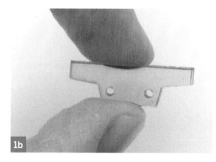

1b

Download CDR or DMG files for the laser cutter, or the print-scale PDF for hand cutting, at makeprojects. com/v/31.

⬛ **TIPS:** If you're hand cutting the fuselage, instead of cutting out the slot for the stabilizer, you can use a saw blade to cut inward from the rear of the fuselage. This is an easier cut, and as effective as the slot on the templates.

The wing pivot is a small piece, so use good-quality hardboard or the layers will come apart.

Smooth the top and bottom edges of the pivot piece with sandpaper; otherwise the sharp square edges will tend to wear away at the balsa wood.

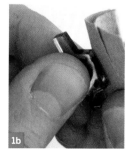

1b

Rick Schertle

1c. For the wing clips, use tinsnips to carefully cut the aluminum sheet into 2 pieces ⅞"×2½". If you're using a soda can, you can use scissors to cut and flatten it out, then trim it to the same dimensions.

1d. Cut the ⅛" dowel to ½" for the wing stop, and the ¼" dowel to 6" for the handheld catapult. For the rubber band tube, cut a ½" length of the rigid plastic tubing. A razor blade will make a clean cut.

2. MAKE THE WING CLIPS

2a. To shape each wing clip, fold the aluminum piece in half lengthwise over a scrap of ¹⁄₁₆" balsa. A vise is helpful in starting the fold. Fold carefully so as to not split the aluminum on the fold.

Using a rotary tool, cut a notch about ⅛" wide and ³⁄₁₆" deep into the folded side of the aluminum clip, ¾" from one end, to accommodate the wing pivot. Debur the notch with a small file.

TIPS: Mount your rotary tool in a vise for increased stability when cutting.

2b. To mount the wing clips, apply a tiny drop of Gorilla Glue to the top and bottom of the wing edge nearest the fuselage, and then slide the aluminum wing clip onto the wing end, leaving enough room for the wire rod to slide freely between clip and wing.

Once the glue dries, clear out the exposed balsa wood in the aluminum notch using the rotary tool.

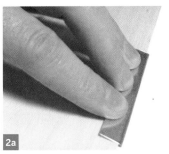

Gorilla Glue works very well, but remember that it expands, so be sure to use only a tiny bit.

Gunther Kirsch (1d)

3. ASSEMBLE THE FUSELAGE

3a. Cut the slot for the nose weight using a power scroll saw or coping saw.

Slip the penny into the nose slot and either glue the penny in place or dip the nose in PlastiDip. PlastiDip gives the nose a bit more durability for hard-surface landings, and looks nice as well.

3b. Use the rotary cutting wheel to cut a shallow groove in the top of the fuselage at the rear, sized to accept the tail fin.

Glue the horizontal stabilizer and tail fin into the fuselage with white glue. Make sure they're aligned, and then add a line of glue on all sides where the tail fin and stabilizer meet up with the fuselage.

3c. Glue the rubber band tube into the front hole in the top of the fuselage, spaced evenly on both sides. A tiny drop of Gorilla Glue adheres well to both plastic and wood.

Glue the wing stop (the ⅛" dowel) into the rear hole in the top of the fuselage with a tiny drop of Gorilla Glue as well. This piece will stop the wings' forward motion when they unfold.

3d. Apply several layers of glue to the pivot hole and the launch notch, letting them dry between coats. This will strengthen the balsa wood, as these areas tend to wear during use.

3a

3a

3a

3b

3b

3c

3d

3d

4. MOUNT THE WINGS

4a. The straddle wire holds the 2 wings together and straddles the fuselage. Cut the wire to about 6½". A large paper clip might work if you unkink it completely and bend it carefully.

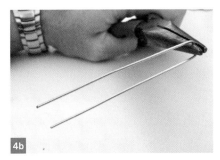

4b. Hold the wire in the center with needlenose pliers, and bend it over the pliers to create a U shape. Then hold the bent wire with pliers ⅞" from the top of the U and bend both legs upward at a 90° angle.

NOTE: This U-shaped saddle will straddle the fuselage and stop the wings' downward rotation into the horizontal position. The straddle wire can later be bent to adjust the angle of incidence in flight. See Use It section.

4c. Attaching the wings to the fuselage is the trickiest step and involves about 6 hands doing different things. Relax and be patient. You'll eventually get everything together correctly.

Hold the wings in the folded-back position against the fuselage, slide in the pivot, and then insert the straddle wire down the gaps in the aluminum wing clips (between the wing and the clip), passing the wire through the holes in the pivot on both sides.

If one side of your wire is slightly longer, try to get that side into the pivot first.

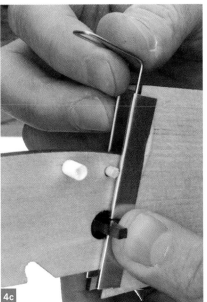

5. ATTACH THE RUBBER BANDS

5a. Turn the glider over and stick staples through each wing, behind the leading edge and 1¾" from the fuselage edge of the wing.

Flip the glider right side up and bend the staples over. Be careful not to tear the thin balsa.

5b. Feed the small rubber band through the tube in the fuselage, then hook it onto the staples and bend them back firmly to secure it.

5a

5c. Loop the large rubber band around the end of the ¼" dowel as shown. This is your handheld catapult.

5b

5d. The glider should unfold its wings in one quick motion. If the wing clips are rubbing the fuselage and preventing the wings from unfolding quickly, spread the straddle wire so that it pushes the wings slightly outward.

5b

That's it! Now you're ready to get outside and launch your very own rocket glider.

◤ TIP: Loop the rubber band around the very tip of the dowel to minimize the chance of the glider hitting the dowel.

◳ Resources

For DWG and CDR files for laser cutting, PDF templates for hand cutting, and video of the Rocket Glider in action, go to makeprojects.com/v/31.

Keep up with the latest instructions and share your build at: makeprojects.com/project/f/1934

More on Jim Walker's Army Interceptor glider: makezine.com/go/interceptor

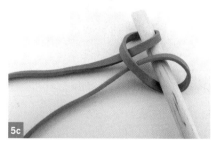

5c

5d

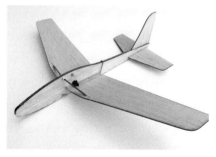

⚑ TEST BUILDER:
Eric Chu,
MAKE Labs

USE IT.

LAUNCH AND SOAR

⚠ Safety Precautions

To launch your glider, select a large ball field or open area away from trees. Grass is ideal, as landings on concrete or dirt tend to eventually eat away at the balsa wood.

As with any catapulted object, use caution when launching your glider. Always point the glider straight up. Never aim it at a person or thing.

Launch Process

» Flip the wings up, from horizontal to vertical.
» Fold the wings back along the fuselage, so the wing tips are above the horizontal stabilizer.
» Hook the catapult rubber band through the launch notch in the bottom of the fuselage.
» Hold the catapult with one hand, and the folded wing tips with the other hand.
» Pull the glider back on the rubber band and launch it straight up!

Launching Tips

Try doing a glide test on soft grass to adjust your glider for a good glide before launching it high. To adjust the angle of incidence, carefully bend the straddle wire forward if the descent is a series of dips and stalls, or backward if the glide is too steep.

It's very important to launch the glider straight up, so that it stalls and allows the wings to open. If it doesn't stall, it may just follow a curve and plow into the ground.

The handheld launcher is ideal for ages 10 and up; you need longer arms to pull the glider back and get it up high. The higher you launch it, the better. This gives the wings plenty of time to unfold for a gentle glide down.

The glider is quite fragile, so caution kids running to retrieve it not to step on it in their enthusiasm! ◢

➕ Stay tuned: In MAKE Volume 32 we'll build a catapult launcher to send this glider even higher!

FETCH-O-MATIC
Build your own automatic tennis ball launcher for dogs.

By Dean Segovis

Several years ago I watched a viral YouTube video that starred Jerry the Dachshund, whose engineer owner had built him his very own automated ball launcher. I had two dogs at the time, and was also unemployed with some time on my hands, so I decided to try my hand at building one.

After a few days rummaging through some junk boxes, I hacked together a slingshot-style automatic ball launcher that actually worked! It was pretty busy with parts though, and nearly 5 feet long. I wanted to go simpler.

Then on Discovery Channel's *Prototype This* I saw a small spring mechanism that I just *knew* would work in larger form for my launcher. It was based on a gearmotor that rotated an offset peg on a wheel. The peg pushed a whacker rod around the wheel and against a spring, until it reached a point where the rod could spring back freely the other way, whacking the ball. I had to build it, and

build it I did. Now you can too.

The Fetch-O-Matic is the third and best version yet of this configuration. It will launch a tennis ball through the air about 25 feet with enough velocity to bounce and roll on for another 20–30 feet. It runs on 12–18 volts DC, so cordless drill batteries are an ideal rechargeable power source.

Dean "Dino" Segovis is a lifelong hacker, tinkerer, inventor, and mentor who enjoys taking things apart and repurposing them. Many jobs over the years have enabled Dino to become a jack-of-all-trades with a wide variety of skills. He currently makes a living as a European automotive technician, and in his spare time documents his weekly projects at hackaweek.com.

SET UP: p.101   **MAKE IT: p.102**   **USE IT: p.110**

HOUND ENGINEERING PRACTICE

The Fetch-O-Matic automatic ball launcher uses a motorized wheel to pull back and release a spring-tensioned whacker arm. Dropping a ball in the hopper clicks a switch to start the motor turning. When the ball has been whacked away, the switch unclicks, and the motor stops.

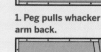

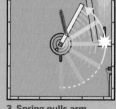

1. Peg pulls whacker arm back.

2. Arm reaches midpoint in back, ready to release.

3. Spring pulls arm forward, whacking ball.

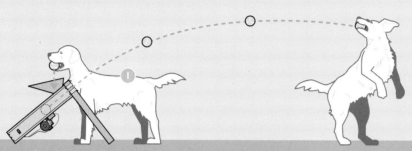

(A) A dog, having fetched a ball, drops it into the hopper.

(B) The ball comes to rest at the bottom of the hopper and its weight closes a micro switch.

(C) The switch completes the circuit, supplying battery power to a low-RPM, high-torque gearmotor.

(D) The motor turns the wheel.

(E) A peg is mounted on the wheel, parallel to and offset from the axle.

(F) The whacker arm rotates loosely around the wheel axle. As the wheel rotates, the peg drives the arm around from one side.

(G) A spring pulls the whacker in the launch direction. The peg pushes the arm back until it's directly in line with the spring, at which point the arm whips forward ahead of the peg, releasing the spring tension and striking the ball. Whack!

(H) As the ball leaves the chute, the switch opens and the motor coasts to a stop, ready for the next launch.

(I) The excited dog hears the whack and runs to retrieve the flying ball.

(J) An adjustable leg lets you angle the Fetch-O-Matic up and down so the ball will follow different trajectories.

(K) A 12V–18V DC battery supplies power to the motor, protected by a rectifier diode.

(L) A power switch turns the launcher on and off.

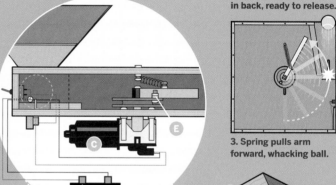

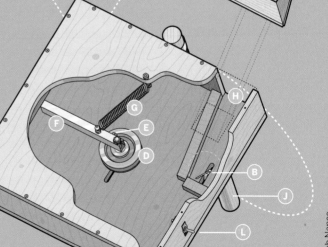

Rob Nance

SET UP.

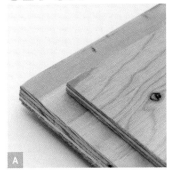

A

B

C

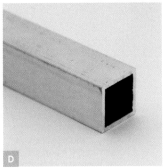

D

E

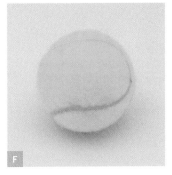

F

MATERIALS

A. Plywood:
» ½" birch or oak, 2'×4', 5 ply minimum
» ¼" birch or oak, 2'×2'

B. Gearmotor, 12V, windshield wiper manufacturer part #F78U-17B571-AA. Online sources list this as fitting a 1996–1999 Ford Windstar.

C. V-belt pulley, 4" with ½" bore Grainger part #3X909, grainger.com

D. Aluminum square tube, ¾", 8¾" length

E. Extension spring, 3¼"×⅝"×0.072"

F. Tennis ball, regulation No substitutions.

» Wood dowel, 1¼"×20"
» Steel spacer for ¼" screw, unthreaded, round, ½" OD × 1½" long McMaster-Carr #92415A871, mcmaster.com
» Battery and charger, 12V–18V
» Switch, SPST, on-off toggle All Electronics #STS-56, allelectronics.com

» Switch, SPDT snap-action micro All Electronics #SMS-196
» Rectifier diode, 1N5406 All Electronics #1N5406
» Hookup wire, 18–22 gauge, stranded: red (3'), black (3')
» Blade connectors: ⅛" (or slightly larger) female (2), ¼" female (2)
» Bolts, ¼-20: 1¼" long (3); 1" long, round head (2)
» Bolt, ⁵⁄₁₆-18, 1" long
» Bolt, M6, 6mm, 1mm/thread × 60mm long
» Lock washer, 6mm
» Washers: ¼" ID × ¾" OD (9), ½" OD (6)
» Nuts: ¼-20 (4), ⁵⁄₁₆-18 (2)
» Wood screws: ½" (4), #6×1½" (50)
» Metal screws, #4×¾" (5)
» Conduit straps, 1" (2) or pipe strap
» Nails, finishing, ¾" (4)
» Cloth strips, 1"×8" (4)
» Wood glue
» Thread-locker fluid such as Loctite
» Masking tape
» Heat-shrink tubing

TOOLS

» Circular saw or table saw
» Jigsaw
» Hacksaw
» Tap and holder, ¼-20
» Drill with guide block, or drill press
» Drill bits: ³⁄₃₂", ⅛", ¼", ⁵⁄₁₆", ⁷⁄₁₆" spade, ½" spade, ¾" spade, 1½" spade, 1¾" hole saw, countersink
» Center punch
» Screwdrivers, Phillips, small and medium sizes
» Ball-end Allen wrench, long reach, ⁵⁄₃₂"
» Wrenches: ⁷⁄₁₆", ½"
» Socket wrench, ⅜" drive, with sockets: ⁷⁄₁₆", 10mm
» Metal file
» Tape measure
» Carpenter's square
» Pencil
» Hobby knife
» Scissors
» Diagonal cut pliers
» Wire strippers
» Soldering iron and solder
» Sandpaper, 120 grit
» Small container of water

Gunther Kirsch

MAKE IT.

BUILD YOUR FETCH-O-MATIC

Time: 8 Hours

Complexity: Moderate

1. CUT THE WOOD PIECES

1a. Cut the ½" plywood as shown in *wood_cutting_guide.pdf*. Start by cutting the length of the 4' sheet and cutting out the two 20" squares.

1b. Mark points ¼" in from all 4 sides of the top and bottom panels, 2", 6", 10", 14", and 18" from either edge. Drill a ⅛" hole at each mark.

1c. Decide on a top and bottom panel. Mark the center of the bottom piece and follow *lower_motor_mount_template. pdf* to mark the 3 surrounding holes. Or print the template, cut out the hole centers, and align and tape it over the board's centerpoint, to mark the other 3 hole centers.

1d. Center-punch all 4 hole centers. Drill the large hole with a 1¾" hole saw. Use a ⁷⁄₁₆" spade bit for the one behind it, and a ¼" bit for the others.

1e. Following *bottom_board_template.pdf*, mark 2 points 7" and 9" from the front, and 1²⁷⁄₃₂" from the left edge (just under 1⅞"). Center-punch and drill with a ⁷⁄₁₆" bit.

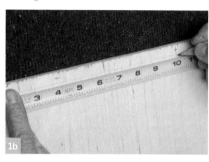

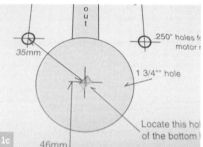

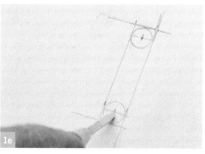

Download and print the 6 project templates at makeprojects.com/v/31.

▧ **TIPS:** With the hole saw and spade bits, drill only halfway through one side, then flip the board over and use the center hole as a guide to finish drilling from the other side. This makes for a cleaner hole with no splintering.

Drill as perpendicular to the surface as possible. For best results, use a drill guide block or a drill press.

Dean Segovis

1f. Use a jigsaw to cut slots between the ⁷⁄₁₆" hole and the 1¾" hole in the middle of the panel, and between the two ⁷⁄₁₆" holes near the left edge.

1g. Follow *top_board_template. pdf* to prep the top panel. Start the 3" square cutout by drilling a hole inside that the jigsaw blade can fit into.

1h. For the right side panel, use a 1½" spade bit to drill a hole centered 5½" from one end. Drill only partway through, leaving 2 or 3 ply layers in place. Then drill all the way through with a ½" bit.

1i. In the back panel, drill a ¼" hole in the lower right corner, ¾" from the bottom and the right side.

1j. Follow *microswitch_ template_detail.pdf* to drill 3 holes through the 1"×2" micro switch mount block.

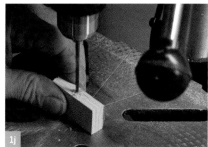

1k. Trace the hopper pieces from *hopper_template.pdf* onto ¼" plywood. Cut them out and drill them with ⅛" holes as shown.

1l. Sand smooth the edges and openings on all the wood pieces.

2. PREP THE MOTOR AND PULLEY

2a. Clamp the gearmotor in the vise and use a ¼-20 tap to tap threads in the 3 mounting bosses of the gearbox.

NOTE: The first motor I used needed approximately ⅛" trimmed off each mounting boss. If your motor is different from the one shown here, you may need to cut less or more, or none at all.

The motor needs to sit flat against plywood with its shaft poking through a hole. If the bosses interfere, trim them down with a hacksaw.

2b. File the cut faces even and flush with the ribs of the gearbox. File the flange near the motor housing flush to the same height. Then file the driveshaft to extend its flat about ¼" back toward the gearbox. Avoid filing the existing flat surface. Take your time, as this is a critical step.

2c. Drill a 5⁄16" hole through the pulley, halfway out from the middle and opposite the setscrew.

2d. To make the peg, run a 5⁄16-18×1" bolt through the hole from the side with the setscrew, and install 2 nuts tight on the other side.

3. INSTALL THE WHACKER

3a. Following *whacker_arm_template.pdf*, cut and drill an 8¾" length of ¾" square aluminum tubing with a ½" hole completely through and a ¼" hole through just one side.

3b. Screw a ¼" nut onto a ¼-20×1" bolt until it's about ⅛" from the head. Run the bolt through the ¼" hole in the tube, install a second nut inside, and tighten it down.

3c. Attach the motor under the bottom panel by running three ¼-20×1¼" bolts with washers from the inside through the ¼" holes. Start at the back hole, screwing into the boss nearest the motor. Turn the bolt just a few turns before adding the others.

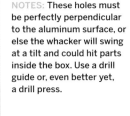

NOTES: These holes must be perfectly perpendicular to the aluminum surface, or else the whacker will swing at a tilt and could hit parts inside the box. Use a drill guide or, even better yet, a drill press.

If your motor shaft is too long, add washers to the top of the motor bosses to lower the motor. The end of the motor shaft should be ¼" above the floor inside the box.

3d. Touch 12V power to the motor until the shaft's flat faces the slot in the bottom panel. Fit the pulley onto the shaft and run the ⁵⁄₃₂" Allen wrench through the slot to tighten the setscrew.

3e. Insert the ½" spacer into the middle of the pulley, then stack 3 or so washers around the spacer until they extend up just beyond the pulley's rim.

3f. Slide the whacker arm onto the spacer and secure it with the 6mm bolt running through a 6mm lock washer and a ¼" washer, tightened just snug for now. Check that the whacker rotates easily around its full arc, hitting the pulley peg on both sides. Check that its midpoint is 1⅜" from the bottom panel, and the shaft bolt has at least ⅝" clearance from the top panel, so the spring will clear it.

3g. Remove the M6 bolt, apply Loctite to it, and reinstall. Remove and reinstall the setscrew with more Loctite. Wiggle the pulley as you wrench to ensure that the screw is centered on the flat. Wrench hard to get the setscrew as tight as you can.

4. ASSEMBLE THE HOPPER & CASE

4a. Join the 4 hopper pieces with masking tape, taping firmly along the inside edges. Use a square to align the assembly. Tape the wide end (the top) down onto a hard

NOTES: Once all the bolts are started, tighten them evenly a little at a time until they feel good and snug, but don't crush the plywood.

Depending on your motor, you may need to use extra washers as spacers to get the spacing inside the box just right.

To supply power to the motor, connect negative to pin 2 (ground; see Step 7a) and positive to pin 3 (low speed).

Gunther Kirsch (2c, 3e–3g)

surface, and tape across the narrow end to reinforce it.

4b. Dampen the 4 cloth strips, wring out, and lay flat. Apply an ample bead of wood glue down a corner joint of the hopper and smear it out about ½" on each side. Apply a strip over each joint, smoothing and working out bubbles. Let dry overnight.

4c. Install the ball guide and backstop as shown on *bottom _board_template.pdf* using wood glue and ¾" finish nails.

NOTES: Go back over all screws and check tightness. They should be flush with the wood surface. You can optionally countersink all the screw holes first.

4d. Starting with the left panel, join the 2 side panels to the bottom panel. Clamp each in place under the bottom, pilot-drill the holes, and install #6×1½" wood screws.

Be sure to position the front panel to the left side, to allow the tennis ball to pop out on the right.

4e. Join the front and back panels to the bottom in the same manner. Finally, join the sides to the front and back with one screw each, centered along the 2¾" dimension.

Sharp-eyed readers will notice that for this magazine cover, we built a left-handed version of the launcher, reversing the templates and switching the motor wires to achieve the opposite rotation.

5. INSTALL THE SWITCHES

5a. Cut the following pieces of hookup wire: 14" red (2), 16" red (1), and 28" black (1). Strip ¼" of insulation from all ends, then twist the strands and tin with solder.

5b. Solder or crimp ¼" female blade connectors onto one end of the 28" black wire, and one end of a 14" red wire. Solder the other end of this red wire to the micro switch's

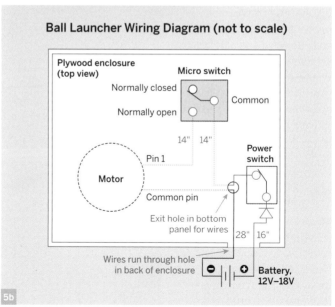

Ball Launcher Wiring Diagram (not to scale)

Plywood enclosure (top view)

Micro switch

Normally closed

Common

Normally open

14" 14"

Power switch

Pin 1

Motor

Common pin

Exit hole in bottom panel for wires

28" 16"

Wires run through hole in back of enclosure

Battery, 12V–18V

normally open (NO) terminal.
Solder the other 14" red wire to the micro switch's common terminal.

5c. Trim the diode's leads to ⅜" and solder the cathode end (with the stripe) to one of the power switch terminals. Solder one end of the 16" red wire to the other (anode) side of the diode, insulating with heat-shrink tubing.

5d. Mount the power switch through the ½" hole and 1½" indentation in the right side panel, with the On side of the switch pointing up.

5e. Pilot-drill and screw the micro switch mounting block under the bottom panel alongside the ball guide hole. Angle the ball launcher upward, roll a tennis ball down the chute, and see where it hits the backstop.
Find a mounting position for the micro switch on its block, angled so that the ball pushes down the roller at the end of the lever. Adjust its position until the ball triggers it consistently, then mount it with two #4×¾" screws.

5f. Route the micro switch's common wire and the 28" black wire through the ¼" hole near the power switch. Pull through a length of black wire equal to the length of the red. Route the rest of the black wire and the power switch's 16" red wire out the ¼" hole in the back panel.
Finally, solder the micro

Micro switch

Power switch

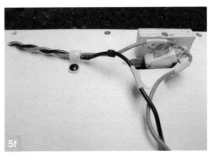

NOTE: Make sure the wires and switch are clear of the arc of the whacker. I used nylon cable guides and zip ties to tidy up the wires.

Gunther Kirsch (5b–5d, 5f)

switch's common wire to the other leg of the power switch.

6. INSTALL SPRING, HOPPER, & LEG

6a.

6a. Install a 1" bolt up through the hole in the top panel, with a nut on each side. Leave a gap between the bolt head and nut. With the whacker pointing forward, hook the spring between the bolt on the top panel and the bolt on the whacker.

NOTE: Test by lowering the top panel and positioning it on the box; it should pull on the spring.

6b. Clamp the top panel in place and screw it to the side panels like you did with the bottom panel in Step 4d, starting with the front right side.

6b.

6c. Place the hopper over the hole in the top panel, with its short side toward the front of the box. Drill four $3/32$" holes through the holes in the hopper and into the top panel, each at a 45° angle except for the right side, which should be drilled 90° straight down into the right panel. Install a #4×¾" screw in each, and tighten in turn just until contact is made with the hopper.

6c.

TIP: To prevent the dowel from rolling while you drill it, clamp it between 2 blocks of scrap wood.

6d. Cut a 20" length of 1¼" wood dowel. Mark and drill centered ¼" holes in-line at these distances from one end: 3½", 5½", 7½", and 9½".

6d.

NOTE: The 1" pipe straps fit the 1¼" dowel; they're actually bigger than 1".

6e. Mark a centerline down the front panel and 2 more lines offset 1³⁄₁₆" on either side. Hold each pipe strap over the 2 outer lines, mark their hole locations, and drill

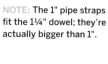

6e.

Gunther Kirsch (6a, 6b, 6e)

with a ³/₃₂" bit. With everything straight and level, screw the straps in place using ½" wood screws.

6f. Slide the dowel through the straps with its holes facing out, and insert a ¼"×1¼" bolt in any hole below the brackets. The adjustable leg is now ready to support weight.

NOTE: From left to right, the motor's first 3 pins are high speed (1), common (2), and low speed (3).

To spin the motor clockwise, switch polarity on the battery.

7. POWER UP AND TEST-FIRE

7a. Turn the case over so the motor's electrical connector faces you. Plug the red wire from the micro switch's NO terminal onto pin 1, and plug the unconnected black wire onto pin 2.

7b. *Turn the power switch off!* Connect its red wire to the battery's positive side, and the black wire to negative.

7c. Set up the Fetch-O-Matic in a flat area with about 40' of space in front. Adjust the leg to a medium height. Make sure the power is still off, and check for foreign objects in the ball chute. Turn the power on, and check that the area directly in front of the launcher is free of dogs, kids, faces, etc.

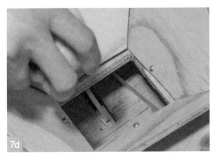

7d. Drop a tennis ball into the hopper. As it drops in and triggers the micro switch, the motor will turn and load the spring. When the whacker is pushed past the center point, it's free to rotate and strike the ball, thus launching it!

TEST BUILDER:
Eric Chu,
MAKE Labs

Gunther Kirsch (6f, 7c, 7d); Gregory Hayes (final)

USE IT.

FETCH ME THE FETCH-O-MATIC

Safety

The velocity of the ball when launched isn't fast enough to do much harm. Holding a hand in front of the unit during a launch will give you an idea of its speed. It doesn't hurt, it just hits a bit hard. Catching one on the tip of the nose won't make Rover happy, but chances are he'll just shake it off and be a bit more cautious next time. However:

⚠ **CAUTION:** The launch arm inside the case *will really hurt you* if it strikes you during release! I got hit once in the finger and it felt like a hammer. It certainly has the potential to damage flesh, so observe these simple safety rules when using the Fetch-O-Matic:

» Keep your hands out of the ball loading area and chute.

» If the unit jams, disconnect the power and try to free it by jarring or shaking.

» *Do not reach inside* if the spring is cocked. Remove the top and unhook the spring first.

» Supervise animals and children.

That said, use caution and have fun with your Fetch-O-Matic automatic ball launcher!

Training Dogs to Use the Fetch-O-Matic

With a dog that likes to fetch, the trick is getting him to drop the ball *into* the hopper and not on the ground next to it. Try working with your dog's favorite treats. Launch the ball and when Rover comes back with it, say "Hopper" and hold a treat over the hopper. Try to get him to drop the ball into the hopper in order to get the treat.

Repeat this until he gets it. Give him some attaboys and a few tosses from your own arm once in a while. Smile and laugh with your dog, who now loves you even more! ▨

🎥 Go to makeprojects.com/v/31 to watch videos of the Fetch-O-Matic automatic ball launcher build and see it in action. Visit dinofab.com/ball_launcher.html for video of Dino's earlier versions.

Gregory Hayes

1+2+3 World's Cheapest Monopod
BY GUS DASSIOS

MOST PEOPLE ARE FAMILIAR WITH
tripods. They have three legs and are great
for setting up at a location and taking photos
from that one spot. Monopods, on the other
hand, have only one leg, so they can be quickly
moved from place to place.

But with only one leg, monopods aren't
very stable. This one has a spike at one
end that can be pushed into the ground for
timed photos. Or you can take fixed-height
orbital photos, which is useful for projects like
Print Your Head in 3D (makeprojects.com/
project/p/2194) in which you take a series of
photos, create a digital mesh online, and then
print out your subject on a 3D printer.

1. Prepare the handle.
If it has a rounded end, cut off or sand the
end so it's flat. Pre-drill each end of the broom
handle: one with the 3/16" bit, the other with
the bit sized for your nail.

Pre-drilling allows the hanger bolt and nail
to go in easier and reduces the risk of the
wood splitting.

2. Attach the hardware.
Secure the hex nut on the hanger bolt, and
then attach both to the broom handle using
a wrench.

On the other end, install the nail by carefully
hammering it into the other pre-drilled hole.
After it's secure, use a hacksaw to cut off the
head of the nail. This will allow the monopod
to be stuck into the ground.

3. Attach the camera.
The final step is to screw on the camera.
Most cameras have a 1/4"-20 threaded hole
on the bottom. Without the threaded hole, it
cannot be mounted to a monopod or a tripod.
If you secure the monopod into the ground,
make sure it's steady before letting go. ▨

Gus Dassios lives, designs, and builds in Toronto, Ontario.

Damien Scogin

YOU WILL NEED

Broom handle or 1"-diameter wood dowel
Electric drill and drill bits: 3/16" and another
   bit narrower than your spiral nail.
Hex nut, 1/4"-20 thread
Hanger bolt, 1/4"-20 thread
Wrench
Spiral nail, 8"
Hammer
Hacksaw
Sandpaper (optional)

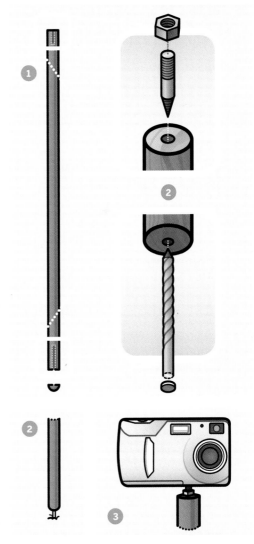

BEGINNER

SERVO CONTROLLERS

Small, dedicated boards drive servomotors without (or with) programming.

By Robert H. Walker

Servomotors do the heavy lifting in countless robotic, animatronic, and other projects, and because they're so visible, it's easy to overlook the role played behind the scenes by their controllers. Your choice of servos for any given project will boil down to the mechanical force you require and the physical size you have room for, and you can buy servos that range from very small to large and powerful. But with servo controllers, the selection criteria are more complex, and include the following questions:

- » How many servomotors will you need to control?
- » Will activation be triggered manually or by an external signal?
- » What's the nature of the triggering signal (continuous, momentary, etc.)?
- » Do you need programmable on-board memory?

Meanwhile, you can also program a general-purpose microcontroller to control servos. For example, Arduino's built-in servo library lets you control 2 servos from your code by connecting their signal wires to I/O pins 9 and 10. This is good for some projects and a nice way to experiment, but for other uses, dedicated servo controllers offer a simpler, smaller, and more elegant solution.

Robert H. Walker has authored numerous articles about model train control and helped pioneer the hobby's use of servomechanisms. He holds a master's degree in electrical engineering and has been awarded five U.S. patents in the field of radio communications.

Photo illustration by Gregory Hayes and Jason Babler

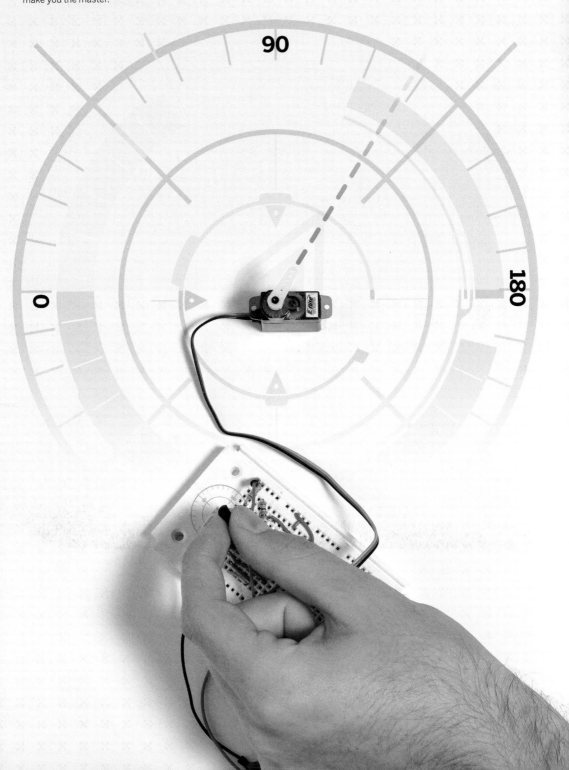

AT YOUR SERVICE
Build one on a breadboard in
minutes, or graduate to multichannel programmable units
— either way, servo controllers
make you the master.

⊞ HOW SERVO-MOTORS WORK

Tod E. Kurt's primer on servomotors in MAKE Volume 19 (page 140) provides a good introduction to how these versatile and inexpensive gearmotors work. To review, you use a 3-wire interface — power, signal, and ground — to make a servo rotate to any position along a 180° radius. Inside the servo, a potentiometer on the motor's driveshaft connects to control electronics that read the driveshaft's current position and move it to where the signal wire tells it to go.

The signal wire conveys its desired position using *pulse-width modulation (PWM)*. Specifically, HIGH pulses are sent every 20 milliseconds (ms), where a 1ms pulse means rotate fully counter-clockwise, 1.5ms means point straight ahead, and 2ms means rotate fully clockwise. If this signal stops, for example if the power is cut off, the motor will de-energize but will typically stay in its last position.

MATERIALS

This article references the following servo controllers, all of which work with standard 5V hobby servos:

DIY 555 manual servo driver components: 555 chip, 50kΩ linear potentiometer, 1N4148 diode, 18kΩ resistor, 680kΩ resistor, 0.033μF capacitor, 0.1μF capacitor, small breadboard, hookup wire

Servo Controller Kit #HH-SCKit from Hansen Hobbies (hansenhobbies.com), $11 each or 3 for $27

Dual Servo Driver #902MSD from ServoCity (servocity.com), $50

Singlet Servo Decoder SSD001, SSD002, or SSD006 from Tam Valley Depot (tamvalleydepot.com), $20 assembled/$12 kit

Quad-Pic Servo Decoder QPC001 from Tam Valley Depot, $38; configured for model train barrel loader with: 4 servomotors, 4 micro switches, toggle switch (for power), two 100Ω resistors, 4.7F supercapacitor, and a 12V SPDT relay

Micro Maestro 6-Channel USB Servo Controller #1350 from Pololu (pololu.com), $20

A

MANUAL CONTROLLERS

Let's begin with the simplest controller: with a **single-driver manual controller**, you control servo actuation manually by rotating a potentiometer knob, and the servo follows the knob's rotation. This driver is useful for simple automation where you can see the mechanism involved and need to control the servo's direction and speed by hand.

A typical example would be a servo mounted in the cab of a model steam shovel and configured so that the shovel moves up and down as you turn the knob (Figure A). A single-driver manual controller can also serve as a quick and easy way to test whether a servo is functioning properly.

You can easily build this basic type of controller out of a 555 timer chip; Figure B shows a typical example, and you can find many variants online. You can do all the wiring in a couple of minutes on a solderless bread-board, or solder it onto a bit of perf board for something smaller and more permanent. Hansen Hobbies sells a kit that's quite similar, based on a 556 dual timer chip (Figure C).

You can use a single manual driver to activate more than one servo by switching its signal line. Figure D shows this circuit configuration with 2 servos, but you can control as many as you want, based on the number of switch positions you have. The main limitation with this arrangement is that you can only

Robert H. Walker

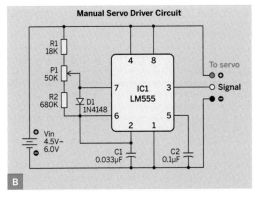

Manual Servo Driver Circuit

R1 18K
P1 50K
R2 680K
D1 1N4148

IC1 LM555

4   8
7   3
6   5
2   1

To servo
○ +
○ Signal
● −

Vin 4.5V–6.0V

C1 0.033µF   C2 0.1µF

B

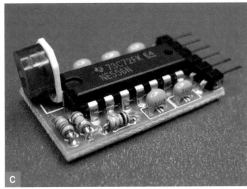

C

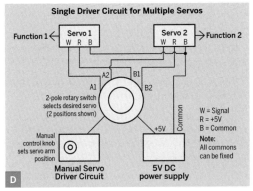

Single Driver Circuit for Multiple Servos

Function 1 ← Servo 1 W R B
Servo 2 W R B → Function 2

A2   B1
A1   B2

2-pole rotary switch selects desired servo (2 positions shown)

+5V   Common

W = Signal
R = +5V
B = Common

Note: All commons can be fixed

Manual control knob sets servo arm position

Manual Servo Driver Circuit

5V DC power supply

D

E

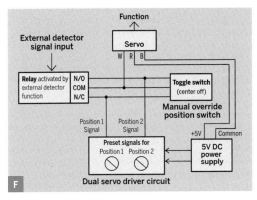

External detector signal input

Function ↑

Servo
W R B

Relay activated by external detector function
N/O
COM
N/C

Toggle switch (center off)

Manual override position switch

Position 1 Signal   Position 2 Signal

+5V   Common

Preset signals for
Position 1   Position 2

5V DC power supply

Dual servo driver circuit

F

G

activate one servo at a time.

Staying with the manual style of activation, **dual-driver controllers** like the Robotzone 902MSD from ServoCity (Figure E) control 2 independent servo drivers simultaneously from one board. Obviously this lets you manually operate 2 servos at a time, but you can also configure the 2 outputs to control a single servomotor in ways that aren't possible with a single-output driver. For example, you can set driver 1 for servo position A and driver 2 for servo position B, then switch the servo

signal lead between them. This moves a servo from A to B in a way that mimics the action of a solenoid, but is smoother, more controllable, and more realistic when used in props.

The servo signal lead can be switched manually or by relay, which opens up as many control possibilities as there are devices that can operate a relay. Figure F, for example, diagrams a circuit that moves a servo between 2 preset positions based on input from either a sensor (such as an IR photodetector) or a toggle switch.

If you want 2 servomotors to simultaneously make the exact same movements, you can connect their signal inputs to share the same driver output. The signal will still work if you split it; you don't need a dedicated driver for each motor. Figure G shows a typical

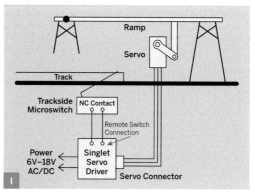

application of this: a model railroad transfer table on which 2 motors move a length of track (with a train on it) back and forth from one set of tracks to another.

PROGRAMMABLE CONTROLLERS

At the next level of servo driver complexity, programmable controllers store a single servo action (defined by a start position, end position, and speed), and then trigger that action automatically. A switch on the board lets you set the trigger mode to respond to either *momentary* or *continuous-on* input.

With momentary, which is more frequently used with pushbutton or sensor input, the servo automatically returns to its start position after moving. A magnetic reed relay sensor detecting a train car passing by is a typical example, or input from an IR detector or inductive current sensor. With continuous

triggering, like from a flipped toggle or slider switch, the servo remains in its end position until the input turns off.

To program the servo's start, end, and speed settings, you follow a programming sequence using small buttons onboard the driver. The specific sequence you need to follow depends on the driver board you're programming, but it's generally a simple process.

The Singlet controller from Tam Valley Depot (Figure H) is a basic **programmable single-driver controller** with a small footprint (1¼"×1¼") that makes it easy to hide. Figure I shows how to configure it to support an automatic culvert unloader triggered by a long-lever micro switch mounted sticking up alongside a model train track. When a train car backs up to the ramp, its body closes the switch. This initiates the servo to tilt the ramp upward and unload the cargo into the car (Figure J).

For more complex projects, we move up to a **programmable 4-channel controller** like the Quad-Pic from Tam Valley Depot (Figure K). Measuring 3"×2.75", this unit provides 4 servomotor drivers that you can program independently, and by configuring them to trigger each other, you can make them perform coordinated routines.

Figure L shows a circuit that uses this controller for a model-loading platform in which a series of servo-operated paddles push barrels up a ramp. The ends of the servo horns opposite the paddles click 3 micro switches (SW1, SW2, SW3) in sequence, in each case actuating the next servo in line until

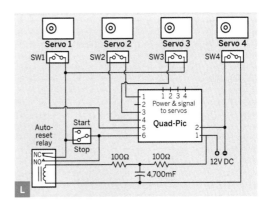

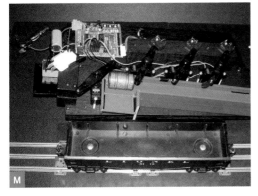

all 4 servos have moved to their maximum clockwise position. The fourth servo then closes the last micro switch, sending power to the auto-reset relay. This causes the cycle to repeat, after a short delay determined by the value of the supercapacitor. Thus, a full load of barrels is conveyed one-by-one up the ramp (Figure M).

COMPUTER-PROGRAMMABLE CONTROLLERS

For the greatest flexibility and complexity, **computer-programmable multichannel controllers** like the Maestro USB series let you write control scripts that define any sequence of actions that you want each servo to perform. The 6-driver, postage-stamp-sized Micro Maestro (Figure N) could control a barrel-loading sequence like the one above without needing any switches. Just trigger it, and the Maestro follows through with its sequence of commands for all the servos. Larger Mini Maestro controllers can handle 12, 18, or 24 individual drivers.

The Maestro drivers make servo control invisible, which is cleaner but possibly less fun to watch, so it depends on what you want. You program the sequences on a PC using the Maestro software, then connect your computer to the Maestro board to upload your control scripts — the same basic procedure as with an Arduino. But these special-purpose servo driver boards let you control more servos from smaller hardware than an Arduino, and use a simple, servo-specific scripting language. I built my original barrel loader using a Micro Maestro, but then decided to go old school with the switches and Quad-Pic because many model train enthusiasts prefer it that way.

Your choice of servo controller is just as important as your choice of servomotor. You have to consider the nature of the signals that you want to initiate the servo actions, and how many servos you need to control. There aren't as many servo controllers to choose from as there are servos, but even with the relatively limited number, there will still be at least one that's right for your project. ◪

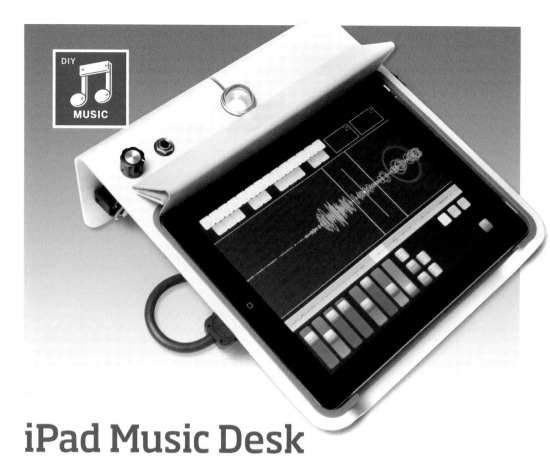

iPad Music Desk

Turn an iPad into a studio-worthy, deep-end experimental musical instrument.

By Reed Ghazala

WHILE I'M BEST KNOWN FOR CHANCE-
based "circuit bending," I've been composing and playing deterministic music for much longer, and I'm always interested in experimental music technology. My hacked iPad 2, broken out into a Music Desk crammed with extreme audio apps, is my favorite new instrument.

On its own, the iPad sounds fine and its large touchscreen is a fantastic musician's interface. But for studio use, the iPad is too lightweight; it needs added stability, better ergonomics, and a more professional wiring interface. That's what this hack is all about.

Here's how you can make a dockable Music Desk for your iPad by adding ¼" line inputs and outputs (Apple has the inputs reserved, but you'll be ready if Apple opens them),

video output, guitar/microphone input with level control, headphone output with level control, a power indicator LED, and a USB jack that will allow both stereo line input and MIDI I/O, via a MIDI-USB adapter.

1. Cut the case.

Following the cutting templates (Figures A and B), measure and mark the hole positions on the Macally ViewStand. Use a center punch to indent all the marks, and then pilot-drill them with a ⅛" bit.

Next, use a step bit (or regular twist bits of the correct sizes), to carefully redrill all holes up to the correct diameter (Figure C). For the ¾" hole to fit the panel-mount USB jack, you can use a hole saw with a metal-cutting blade

Reed Ghazala: Curtis app by The Strange Agency

MATERIALS

See makeprojects.com/v/31 for full-sized schematic diagrams and cutting templates. Also, check Kineteka (kineteka.com) for a possible kit for this project.

ViewStand aluminum viewing stand for iPad
from Macally (macally.com), item #VIEWSTAND
PodBreakout Mini or PodBreakout Nano Style 1
from Kineteka (kineteka.com), items #POD-DOCK-MINI or # POD-B-N-S1
Potentiometers, audio taper, panel mount:
50kΩ (1) and 1kΩ stereo (1)
RCA video jack, panel mount
Audio jacks, ¼", panel mount: mono
(4 unswitched and 1 switched) and stereo (1)
USB jack, Mini-B type, panel mount I used Bulgin #PX0446.
LED, blue, 5mm
LED housing I used a vintage lens from my collection, but you can also use a ping pong ball (see Step 2).
Knobs, to fit ¼" shaft (2)
Cable clips for 5mm cable, adhesive (4)
Hookup wire, stranded, insulated copper,
22 gauge or so
Centronics parallel printer cable These cheap, outmoded cables pile up at second-hand shops.
A/V cable, ⅛" (3.5mm), 4-pole These have 4 rings on the plug.
Heat-shrink tubing, ⅛" diameter

TOOLS

Ruler
Pencil
Center punch and small hammer or mallet
Drill press (preferred) or hand drill
Drill bits: ⅛", step bit
Hole saw, ¾" or a reamer to enlarge a smaller hole to ¾"
Adjustable wrenches or nut drivers, small
Soldering pencil with chisel tip, 1mm–2mm
Solder
X-Acto knife
Wire strippers and cutters
Multimeter
Lighter or heat gun for the heat-shrink tubing

or use a reamer to expand the largest hole you can drill.

2. Mount the components.

Place all the jacks and pots in their proper holes and hand-tighten their mounting hardware. The stereo pot is for the headphones. Using nut drivers or small adjustable wrenches, tighten all the hex nuts.

The ViewStand has a large hole in the middle for a plastic stabilizer that prevents it from tipping forward. This is where you install

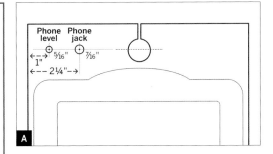

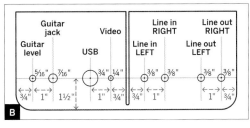

the LED. I enclosed mine in a vintage Cold War lamp housing that I adapted (Figure D), but you could also drill a ping-pong ball, wire the LED inside the ball, and glue it behind the opening. Be creative and make the dock your own!

Want to save battery power? Add a switch

for the LED, or find something else to do with that big round hole. A vintage analog mini VU meter would be cool. So would a nixie tube or magic eye radio tube winking with the music. With the right amp and limiters, both options are possible.

3. Wire the components.

Follow the schematic diagram (Figure E) to wire the components, soldering the connections with hookup wire. Start with all the grounding wires; I used green for these. Then solder the potentiometers (yellow wires): one wire between the guitar pot and "hot" side of the guitar jack; 2 wires between the "hot" sides of the line output jacks (note R/L) and the headphone pot; and 2 more from this pot into the stereo headphone jack (Figure F; cable clips shown are added later).

Remember that LEDs are polarized. The cathode (the shorter leg, on the side with the flat face) should connect to ground.

4. Make the docking cable.

It's time to create the 30-pin docking patch cable. Referring to Figure F, cut a 22" section from your printer cable. Strip 1" of insulation from one end and 8" from the other (lightly score the outer plastic with a sharp knife and then pull it apart). Leave the inner foil shield intact.

Trim the foil shielding away from the 1" wires. From the other end, where the outer insulation stops, pull the bare ground wire and 4 other wires through the foil. Pull another 7 wires through the foil 3" further down the cable. Loop the remaining foiled wires and reserve them for later expansions (Figure G).

At the 1" end, noting color codes, cut away all the wires that you did not pull through the foil at the other end (those reserved and still wrapped in foil). This will leave 12 wires at the 1" end including the bare ground wire.

The next step is to solder the 12 wires to the Kineteka PodBreakout Mini, a finely crafted experimenter's board with soldering holes connected to a 30-pin docking plug. You'll connect all but 3 of these wires to

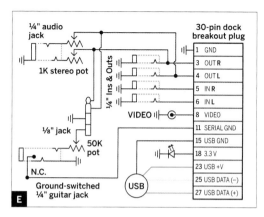

E

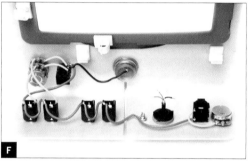

F

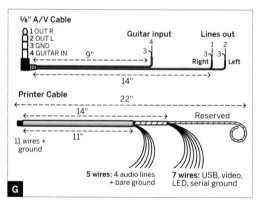

G

NOTE: Be careful when sorting colors; lookalike colors will have a white stripe, or something else to differentiate them, though the difference is sometimes easy to overlook.

soldering holes, which makes this an easy job, so long as you trim the wires to the approximate lengths needed. The remaining 3 wires connect to pins on the PCB, which is also easy with a fine-point soldering pencil if you twist and tin the wires prior to soldering. (If you're using the new PodBreakout Nano Style 1, follow the schematic but solder the 3 "pin" wires into their respective pads.)

Follow Figure H to assign wires to functions and contacts on the PodBreakout. The color-coding scheme is up to you, but be sure to chart it as you go along. For example: Pin 3, Right Out = red wire; Pin 4, Left Out = orange wire with white stripe; etc.

Holding the 1" end of the cable against the PodBreakout with the insulation inside the collar, snip, strip, and tin the 12 wires to approximate length according to destination. Give each one an extra ⅛" for soldering. Leave 3 of the insulated wires a little longer; one must reach to the backside and 2 will run to pins a little past the soldering holes on the front (Figure I).

Starting at one end, connect the 9 shorter wires to their holes by soldering their tips sticking out the backside of the board. The bare ground wire goes to hole 1, and the 4 other wires that you first tugged through the foil at the other end of the cable are for the Line In/Out jacks and go to holes 3, 4, 5, and 6.

Solder the 3 longer wires to pins 11 and 15 on the front, and pin 18 on the back, running the wire through empty hole 21 (Figures J and K). To solder to the pins, hold the tinned end of your wire along and on top of the pin. Lightly place the soldering tip on top, and remove it as soon as you see the solder of both wire and pin flow together. Hold the wire until the solder hardens.

Double-check the wiring. Be sure you've charted all color-to-plug connections. Align the plug hardware and snap its case together.

5. Make the ⅛" patch cable.

We now breeze on to hacking the 4-pole A/V cable. Cut the phono plugs off about 14" from the ⅛" jack and strip the ends.

Checking continuity with a multimeter, ascertain which rings of the plug terminate in which wires (ring 3, counting from the tip, is ground and will connect to the shields of all 3 cables).

Peel the cable back to 9" and cut to that length the wire that connects to ring 4 of the jack, the guitar input (Figures G and L).

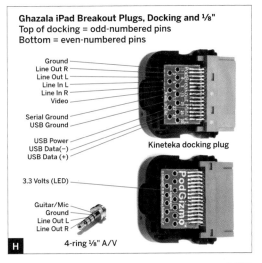

Ghazala iPad Breakout Plugs, Docking and ⅛"
Top of docking = odd-numbered pins
Bottom = even-numbered pins

Ground
Line Out R
Line Out L
Line In L
Line In R
Video
Serial Ground
USB Ground
USB Power
USB Data(−)
USB Data (+)
3.3 Volts (LED)
Guitar/Mic
Ground
Line Out L
Line Out R

Kineteka docking plug

H — 4-ring ⅛" A/V

I

NOTE: Tinning is an ultra-thin application of solder, simple but very important. Get the target hot, and then apply a tiny bit of solder until it coats the metal or is absorbed into a stranded wire.

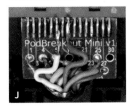

J

K

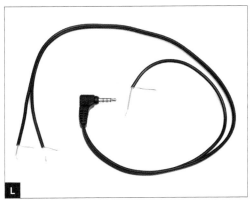

L

6. Connect the cables.

Attach the adhesive cable guides as shown in Figure F, and run the breakout cable through them. Solder the color-coded wires to their destinations, referring to Figures H and M. Trim, strip, and tin as you go, and insulate the USB connections with heat-shrink tubing.

Also run and connect the ⅛" A/V cable (Figure M). Use a X-Acto knife to trim the plug down if it hits the edge of the desk; mine did.

Check your wiring and install the 2 knobs on the pots. Set your iPad in place, plug the PodBreakout into the docking port and the A/V plug into the headphone jack (Figure N). Your iPad Music Desk is complete!

Studio Ready

Using the iPad Music Desk is a pleasure, since it now not only sounds good, it also feels like a serious, professional piece of gear.

» Line outputs allow patching signals to amps, mixers, and so on.

» The USB port allows connection to USB-compliant I/O devices (line input, MIDI, preamps, etc.), however, Apple recommends that you only attach cameras to the iPad via USB. It's up to you to determine whether your USB device will work. As with any hack, you're on your own, Mr. Tesla.

» The Guitar/Mic input provides only adjustable attenuation, so as not to add unexpected components to the signal path. This lets the iPad Music Desk work with devices like the AmpliTube iRig, for example, which plugs into an iPhone or iPad's headphone jack to turn it into a broadly configurable guitar effects pedal. (If you use one, turn your Desk volume all the way up so that iRig "sees" the expected signal path.)

» Finally, the composite video jack provides Apple Slideshow output.

I really dig running experimental apps like Droneo, Curtis, or Reactable into a big sound system, with fingertips on the graphics of the iPad's responsive screen, the whole thing broken out of iPad frailty and now feeling Moog-solid. It makes me think back to the Moog's cool ribbon controller, and the Martenot's

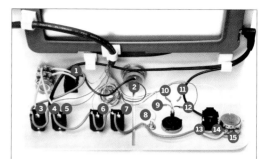

1 To PodBreakout pin 1	**6** To PodBreakout pin 5
2 To PodBreakout pin 18	**7** To PodBreakout pin 6
3 To PodBreakout pin 3 and Line Output (R) from ⅛" jack	**8** To PodBreakout pin 8
	9 To PodBreakout pin 23
	10 To PodBreakout pin 27
4 To grounds of Line Outputs from ⅛" jack (twist right and left ground braids together)	**11** To PodBreakout pin 25
	12 To PodBreakout pin 15
	13 To ground of guitar cable from ⅛" jack
5 To PodBreakout pin 4 and Line Output (L) from ⅛" jack	**14** To PodBreakout pin 11
	15 To center wire of guitar cable from ⅛" jack

M

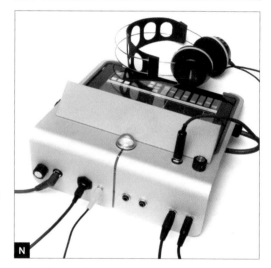

N

weird finger-ring playing system. Wow, music controllers have come a long, long way! ◼

⊞ Download the schematic and templates at makeprojects.com/v/31. See the author's video on pin soldering at makezine.com/go/imdpins.

Qubais Reed Ghazala (anti-theory.com), founder of the circuit bending movement, works as an author, artist, photographer, composer, and teacher. He has designed/consulted for Tom Waits, Peter Gabriel, Blue Man Group, NIN, and Blur, and his work resides in the Whitney, Guggenheim, and MoMA museums. He wrote "Build an Incantor" for MAKE Volume 04.

DIY IMAGING

Monkeysailor's Photo Lab

Make your own Arduino-controlled color film processor.

By Andrew Lewis

TRADITIONAL PHOTOGRAPHY IS
fantastic. I love the way mechanical cameras feel in my hand, and I love the way film makes me think about composition and lighting before I actually take a shot. The only thing I don't like about traditional photography is the cost of having film processed or buying equipment to develop my own.

To cut costs and have a little fun with an Arduino along the way, I decided to make my own film processor. The equipment needed to turn a roll of exposed medium-format color film into negatives actually isn't that complicated, and the chemical process is also straightforward. Using a standard developing tank and chemicals, all you really need are a stable working temperature and accurate timing, which you can accomplish using an Arduino, an LM35 temperature sensor, and a

few electrical components.

To develop film, you immerse it in developer at a specific temperature for a specific amount of time, agitate it every few seconds to ensure an even process across the film, then repeat the process with fixer/blix solution, and repeat again with rinse water. This can be done in a Paterson tank, which lets you pour liquids in and out without exposing the film inside to any light. With a film-changing bag to load the film into the tank, you don't need a darkroom!

Different film types or chemicals require different times and temperatures, and the brightness, contrast, or color will not develop correctly if they're wrong. So I needed to work out a system that would maintain and monitor the temperature of 4 chemical bottles, and time the processing down to the second.

MATERIALS AND TOOLS

Paterson film developing tank, film developing chemicals, and film changing bag

Arduino-compatible microcontroller board
I used a Seeeduino.

PC power supply unit (PSU), ATX standard, 650 watt or similar

Servomotor, 5V, general-purpose hobby type
I used a Futaba S3001.

LCD display, 16×2 character, no backlight
I used a Hitachi HD44780.

Circuit board for display buttons. I etched my own PCB (download the mask at makeprojects.com/v/31). You can also use plain perf board, about 4" square.

Temperature sensors, LM35 (2)

Switches, momentary pushbutton, normally open (NO), 10mm, through-hole, (8)

Transistors, STP36NF06 MOSFET (2) to control the heater elements. You could also use a smaller transistor like a 2N222 to trigger a relay, and a diode to mop up any back EMF from the relay coil.

Resistors: 120Ω (9), 1kΩ (3), and 10kΩ (1)

Potentiometer, 47kΩ linear

Speaker, 8Ω, small

Adapter cable, ATX-24 to ATX-20

Spade or bullet connector pairs (4)

Pin headers, male-male breakaway, 1×40 These make it easier to connect wires to the Arduino securely.

Wire, stranded, 20 gauge or so for signal connections; I used old ribbon cable.

Wire, stranded, 12 gauge or so for power connections; I used multicore Power Flex from an old appliance.

Nichrome wire I used old heater wire that has a resistance of about 2Ω per yard.

Wire, stiff, 12" length

Wire, copper, 36"

Wooden case, ¼" plywood I got one from Hines Design Labs (angushines.com).

Plastic sheet, rigid, about 3mm thick, 185mm × 155mm It won't be visible, so it's OK if it's damaged.

Aluminum sheet, 1⁄16"×2'×2'

Tin can, 4" (100mm) diameter for the Paterson tank warmer. I used a cigar can; a big soup can will also work.

Panel printout on thick paper or printable plastic film. You can design your own or download mine.

Furnace cement aka fire cement or refractory cement, to encase the heater coils

Various fasteners for mounting the warmers

Mesh, fabric or wire to cover the vent

Fiberglass insulation, scraps

Silicone sealant, epoxy resin, and wood glue

Soldering equipment and solder

Band saw for cutting PVC and sheet aluminum

Drill and drill bits suitable for sheet metal

Pop rivet tool and rivets

Wire strippers and cutters

Screwdrivers

Scissors

Glue gun and hot glue

Adjustable clamps (2)

Computer with internet connection and printer

A   SIG GND 5V   B

I also thought it would be handy if my Photo Lab could agitate the film automatically, and store my time and temperature settings so I wouldn't have to reprogram them with each batch. For the auto-agitation, I used a hobby servomotor, and for programming I designed a control panel around a Hitachi HD44780 16×2 LCD display, with pushbuttons for making menu selections. The buttons let you move through the menu screens, increase and decrease heater, timer, and agitation values, and save/restore the settings to the Arduino.

With the agitator, timer, and menu system, the Photo Lab project plans grew larger than my original idea for a chemical warmer, but I was confident I could make it all work. At one point I realized that although I had set my sights on a photographic processor, my system would be great for warming any liquids. With a little modification, it could be used to control hotplates, furnaces, and fish tank heaters for other projects.

1. Build the control panel.

At first, the Arduino I/O pins seemed outnumbered by the connections needed for the LCD screen, temperature sensors, servo, and pushbuttons. To get around this problem, I made a custom PCB with all the buttons wired in parallel to a resistor array with different-value resistances along each path (Figure A). Each button press produces a different voltage through the board, which lets the Arduino read all the buttons from a single analog input pin.

For power, I used a standard ATX power supply unit (PSU) for tower PCs. These

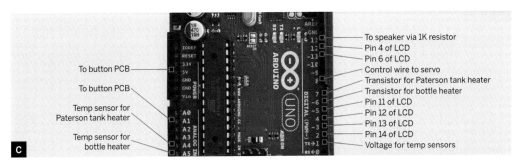

To button PCB

To button PCB

Temp sensor for
Paterson tank heater

Temp sensor for
bottle heater

C

To speaker via 1K resistor
Pin 4 of LCD
Pin 6 of LCD
Control wire to servo
Transistor for Paterson tank heater
Transistor for bottle heater
Pin 11 of LCD
Pin 12 of LCD
Pin 13 of LCD
Pin 14 of LCD
Voltage for temp sensors

provide plenty of amps, have a built-in fan and surge protection circuit, and make available a nice selection of voltages via their 20- or 24-wire cable that connects to the PC motherboard. To turn the power supply on, you connect its green and black (ground) wires, and then the yellow and red wires supply +12V and +5V, respectively.

I wired my circuitry to the PSU via a cut ATX adapter cable (Figure B), which saved me from cutting the PSU's own cable, but you could cut and solder the unit's wires directly. I connected the Arduino to a 12V line. I could have bypassed the Arduino's internal voltage regulator and connected it directly to 5V, but felt that it wasn't necessary.

See Figure C for a schematic diagram of the Photo Lab's microcontroller connections (and for a full schematic of this project, see makeprojects.com/v/31). I glued the button PCB and LCD screen onto a sheet of PVC with holes cut through for the display and buttons. I then made a slightly larger aluminum sheet (195mm×175mm) with matching cuts to the front. I drilled its corners for screw mounting to the main box, and glued the aluminum onto the PVC.

I chose an LCD display without backlighting so that I could use the Photo Lab in a darkroom, and I powered it with 12V from the power supply. Following the LCD's datasheet, I connected a small potentiometer to control the contrast, and just glued it to the back of the panel because once it's set, it doesn't need changing. After I connected the LCD and speaker to the Arduino and glued them to the PVC, the interface hardware was complete (Figure D).

D

A designer friend came up with a nifty retro design for a panel cover, styled after an old camera, which I printed onto a sheet of plastic film. Aside from making the panel look pretty, this cover acts as a splash guard for the push buttons.

Finally, I made sure everything was glued securely in place, and I uploaded my Arduino sketch to the microcontroller for testing. Download it at makeprojects.com/v/31.

2. Connect the temperature sensors.

Now that I had my panel, I could use it to control and sense things in the real world. I experimented with thermistors and thermocouples, but eventually decided to use two LM35 solid-state temperature sensors, one for the chemical bottles and one for the developing tank. These sensors are accurate to about 0.5°C, which is good for sensitive color film processing.

I powered them in parallel from the Arduino's D1 pin, and for greater accuracy, they're only powered up when they're about to take a reading. Continuously supplying them with power generates a bit of heat that can throw their readings off.

3. Mount the servo.

I attached the servomotor to the lid of my developing tank with hot glue, and added a stirrer to the servo shaft using bent wire. To avoid the tank lid being permanently attached to the Photo Lab along with the servo, I added a servo connector to the top (Figure E), which is wired to the main board's power, ground, and Arduino pin D8 for control.

4. Build the bottle warmer.

To make a bottle holder, I measured the 4 bottles that I use to store developing chemicals, and constructed an aluminum box to fit neatly around them. I used thin aluminum sheets, and riveted them to aluminum angles at the corners (Figure F). The pairs of side pieces measured 90mm×150mm and 320mm×180mm, and the bottom piece measured 90mm×320mm.

For the heating elements, I wound Nichrome heating wire around a screwdriver shaft to make 4 coils. I connected these in parallel between 2 copper power wires, and arranged to sit one under each bottle in the bottom of the aluminum box (Figure G).

The coils are powered by the 5V line from the PC power supply, switched by a STP36NF06 transistor controlled by Arduino pin D6. Alternative coil winding and voltages will produce different results, so you can tailor the element design to suit your needs.

I wound the heater wire into coils, and set them into fire cement at the bottom of the aluminum box (Figure H), with the power wires sticking out one end to make wiring them into the control panel easier. I added another sheet of aluminum to close the bottom of the box, then sealed it with silicone sealant for neatness.

I mounted the temperature sensor on the bottom of the aluminum box near the wires that lead to the heating element. The sensor connects to Arduino analog pin A5. I did think about suspending it inside one of the bottles, but I felt that having the sensor on the element would be neater.

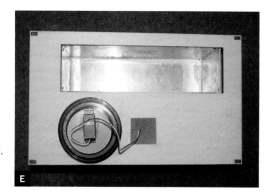

E

F

G

H

5. Build the developing tank warmer.

Paterson developing tanks are cylindrical, so instead of making a holder out of sheet metal, I used an old cigar can that fit it nicely (Figure I). A large soup can also works.

I drilled a hole in each side of the tin and fitted a Nichrome coil in place with fire cement as I'd done with the bottle warmer.

Finally, I glued the temperature sensor to the tin with epoxy resin. This heater coil is controlled by a transistor connected to Arduino pin D7 and the sensor feeds into pin A4.

6. Make the case.

The laser-cut Photo Lab case was manufactured from birch-faced plywood by my friends over at Hines Design Labs. I designed the case to accept an ATX power supply, and before I fitted it in, I covered the air intake with wire mesh (Figure J). You can see my templates at makeprojects.com/v/31.

7. Final assembly.

I connected several of the power supply's 5V (red) lines together so they could handle the current for the heater elements. The power switch for the PSU acts as the power switch for the Photo Lab, and I connected the green and black ATX wires together so that the PSU supplies power as soon as it's plugged in and turned on.

I fit the aluminum box and cigar can into the project box using a combination of wooden blocks, Meccano brackets, wood screws, and wood glue.

Nothing inside the box should get hotter than 140°F (60°C), so I didn't make any special effort to protect the wood. As a final measure, I packed some fiberglass thermal insulation around the heaters.

The Big Picture

I'm very pleased with this project, and I now use it to process my rolls of film. The only problem so far is that the heaters take a long time to bring the bottles up to temperature, so I think I might replace the elements with 12V or 240V versions. This should be easy because it

involves no modifications beyond exchanging the transistors for solid-state relays.

The timer and auto agitator take all the stress out of processing film, and mean that I don't have to keep an eye on the clock. I can leave the film to process and trust that the tank will be stirred every few seconds.

The processor is also handy for developing black and white film using weak solutions to achieve finer grain, which is a slow process. Experimental developers such as caffeine (yes, caffeine!) can take more than half an hour to process, and my Photo Lab machine lets me avoid a nasty hand cramp from manually stirring a tank for 30 minutes. ◪

⊞ Download the Arduino code, schematic diagram, and Arduino connection table at makeprojects.com/v/31, where you'll also find the optional box cutting templates, panel artwork by John Ranford, and pushbutton PCB etch mask.

Andrew Lewis is a keen artificer and research scientist working in archaeometrics and complexity science at the University of Birmington. He is a relentless tinkerer, whose love of science and technology is second only to his love of neo-Victoriana.

Getting Started with Multicopters

Tips on how to build, buy, fly, and spy with multirotor R/C helicopters.

By Frits Lyneborg

A MULTICOPTER IS A FLYING ROBOT resembling a wagon wheel — without the wheel. It has a central hub with electronics, power, and sensors, onto which are mounted arms that hold propellers to provide lift. The number of arms gives the name: a tricopter (trirotor) has 3 arms, a quadrocopter or quadcopter (quadrotor) has 4, a hexacopter 6, and an octocopter 8. There are other variations, but these are the most popular setups.

They're also called *multirotors*, which arguably is the correct term, but I'll stick to *multicopters* because that's used more often on the internet, where you'll find the most information on the topic.

Why try multicopters? Perhaps you saw one and you just have to own this cool new gadget. Or you fly R/C planes and you'd like to try a new type of aircraft. Or you're into DIY electronics or robots, or you want to do aerial photography. Whatever your motivation, there's an option for you. I've flown a variety of multicopters and built 3 of my own, so I've picked up a few tips I can share.

Homebrew Pedigree

In 2003, Hong Kong-based company Silverlit Electronics read in the newspaper about students Daniel Gurdan and Klaus M. Doth's prize-winning entry in Germany's national

Young Scientists competition. Gurdan and Doth's design was of a radio-controlled, self-leveling quadrocopter (Figure A).

In late 2004, Silverlit began production of their X-UFO, a simplified and cheaper version of the students' design (Figure B). When this product hit the international markets over the next few years, it seeded the idea of a small, remote controlled multi-copter to many people throughout the world. Today there are dozens on the market.

A

How They Work

On an ordinary helicopter, the tail rotor provides horizontal thrust to counteract the main rotor's torque, in order to keep the helicopter from spinning around with the main blades (Figure C).

A multicopter works quite differently. Take for instance a quadrocopter: every second propeller spins in opposite directions (Figure D), counteracting each other's torque. More importantly, a multicopter has an onboard computer that varies the speed on individual propellers, making possible every form of spin, tilt, yaw, and rudder control around any center and any axis, as well as flight in any direction.

B

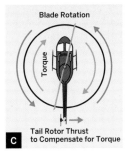

Blade Rotation

Torque

Tail Rotor Thrust
to Compensate for Torque

C

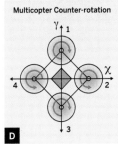

Multicopter Counter-rotation

γ 1

χ

4

2

3

D

Your First Multicopter

The best starter multicopter is lightweight: the lighter the copter, the less damage to it and to the surroundings when you crash. And you will crash! The bigger they get, the more scared you'll be of flying them. Large multicopters can rip through clothes and flesh, and they cost a lot of money.

The downside is that lighter versions carry less payload (read: cameras and extra sensors), and flying time is usually shorter. The upside is that they're cheaper.

Ironically, it's also a good thing that lighter copters are typically harder to control, due to fewer sensors and less-sophisticated overall construction. Why is this good? Because you'll learn to fly. A heavy, complex autonomous multicopter might be easier to fly — or even fly by itself — but you'll never learn

to handle a multicopter that way. That can be a big problem the moment something goes wrong. And something will go wrong.

» QuadPod A nice lightweight starter multicopter is the QuadPod from Snelflight (snelflight.co.uk). They sell a quadcopter kit for £200 ($320) and a complete bundle with batteries and R/C gear for £300 ($475), and they ship worldwide. The QuadPod is a genuine "father and son" project: it takes about an hour to assemble, comes with a nicely illustrated guide, and is very easy and enjoyable to make (Figure E, following page).

In the beginning you'll find it hard to master flying the QuadPod, but it's fun anyway because it's so fast, "live," and light, so you'll learn by playing.

All multicopters are quite sensitive to wind,

but the lighter ones are most sensitive, so the QuadPod is for calm weather or indoor flying only. It's not much bigger than a dinner plate, and weighs almost nothing, so it can withstand crashing on grass. And though you should duck before it hits your face, it won't cause much harm if you do get hit by it.

Though Snelflight says the QuadPod can lift a camera, is hackable, and that they're working on expansions, I recommend you buy it if you just want to learn to fly and have fun — it's not a "serious" multicopter, but a great way to learn to fly.

» Parrot AR.Drone One popular consumer-level multicopter is the Parrot AR.Drone quad-copter (ardrone.parrot.com), with 2 onboard video cameras and its own wi-fi network that lets you control it from Apple and Android smartphones and tablets (Figure F).

I advise you to try out a friend's before you place your order, or buy one used. While these copters are impressive on specifications, they sell for about $300, and at that price you might realize that flying time isn't as long as you'd like, or that it's not as hackable as other machines, and it has little capacity for pay-load beyond its built-in cameras. What it does is definitely cool — but it's not necessarily going to do much more.

» X650 If you insist on having more advanced options right away, and if you dream of lift-ing a camera, perhaps playing around with GPS and more advanced stuff, consider the XAircraft X650 (Figure G). This is a medium-sized quadrocopter that comes as an easy kit.

I think the cheapest way to get one is purchasing directly from China. DFRobot.com has X650s in various configurations for about $600–$750, and a little secret is that they give you 5% off if you use promo code ALABTU when checking out.

The X650 is surprisingly well-built. The details in the finish surpass any kit I've seen, and the manual is even fairly good. It comes as a ready-to-fly (RTF) kit, with a lot of add-on options typically seen on much larger models:

PC link and debugging software, compass, GPS, GoPro camera mounts, fiberglass or car-bon body, and optional 8-motor double blade system (4 double rotors) for extra payload.

Be aware that in this class of hobby aircraft, you're expected to install R/C gear and other necessities yourself. Also, an X650 will need a lot more space to fly than a small backyard.

» Quads or Tricopters? Quadcopters like the X650 and QuadPod maneuver by variously speeding up and slowing down the individual propellers. This is handled by an electronic control board, so all you have to do is learn how to use the control sticks, and watch the magic happen.

Jason Jenkins (E)

This same method is used for all multi-copters with 4 propellers and more. But be sure to look into how tricopters fly. They have a different way of turning, with a servo on one arm, giving an extraordinary cool and futuristic Superman-like flight.

I haven't tried any tricopter kits, but they're available online, and there's a good discussion of the pros and cons in the forums at diydrones.com. David Windestål has posted good videos and build instructions for his Tricopter V2.5 at rcexplorer.se (Figure H).

» Mikrokopter If you're buying your first multicopter, avoid products from Germany-based Mikrokopter. Their products are like a magnet for newbies because of the impressive technical specifications and videos on YouTube, but their products, pricing, and level of documentation and service are by far the worst, especially for people new to the multicopter jungle.

Mikrokopter has a reseller program, so you'll find them in many places. If you really need one, I recommend you get it from mikrokopter.us, because that shop is run by people who try hard to do well where the original German shop does poorly.

Building Your Own Multicopter

Once you've played with multicopters, you'll realize that building one is a project that you could take on. Here are the basics:

» Batteries and Motors The real magic here is the combination of the very powerful lith-ium polymer (LiPo) batteries and brushless motors. These 2 components, with just a nor-mal R/C plane propeller on the motor, can lift themselves right off the ground, and so this combination can make virtually anything fly.

» ESCs and Control Board A multicopter's flight must be controlled and balanced in a certain way. The motors are controlled by little units called *electronic speed controllers (ESCs)*, and these need signals telling them how much power to pass on.

H

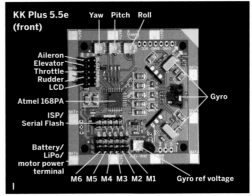

I

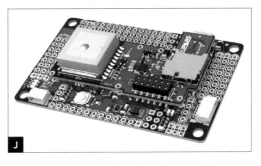

J

In a multicopter, that signal comes from a special control board. The control board is hooked up to a standard R/C plane receiver, and possibly other peripherals such as GPS, or whatever your imagination and wallet allow.

Probably the 2 most popular control boards right now are HobbyKing's Multi-Rotor Control Board V2.1 (hobbyking.com) and Multi RC Shop's KK Plus V5.5e Multicopter Controller (multircshop.com), both based on Atmel's ATmega168 microcontroller chip (Figure I).

Arduino-oriented makers might prefer DIY Drones' ArduCopter system (code.google.com/p/arducopter), with its ArduPilot Mega board based on the ATmega2560 (Figure J).

 Johanna Windestål (H)

» Body and R/C Gear The body of a multi-copter can be made of almost anything, including wood, so the only "mysterious" thing is the control board. The rest is common R/C gear: a 4-channel transmitter and receiver, and connectors to hook up your components.

A Google search on "multicopter control board" will get you started and lead you to plenty of build instructions, and I recommend visiting rcexplorer.se, hobbyking.com, kkmulticopter.com, and diydrones.com.

Video from a Multicopter

Filming from the sky is the most common broken dream among multicopter users. Unfortunately a lot of people are spending a lot of money hoping to make great professional video from the air at a fraction of the cost of a real helicopter. Many shops out there are ready to sell this dream, which I think is unfair.

You should think twice. Here's a test: take your camera and put it on a broomstick. Hold the other end of the broomstick. Now try to get good footage out of that. While it may give interesting new angles and be "arty," in general it's going to look "filmed from the end of a broomstick." You'll find it hard to get the quality of shots you're used to.

The same is the case with a multicopter. You can find cool-looking videos made from multicopters on YouTube, but they're always focused on the flight experience ("Look, I'm flying!"), rather than a specific object or person being photographed.

If you work hard with your equipment, you can get cool shots, but they'll be lucky shots, unless your copter can transmit video back to the ground (see Video Downlinking section). If you get a picture of a house, it'll be awkwardly framed. If you video anything other than random treetops, the subject won't be well placed in the frame, and everything will be moving about. It's not easy.

» Gimbals and Gyros You can purchase very expensive camera mounts and gimbals with gyroscopic stabilization. But before you do, ask to see raw film of at least 1 minute made

⚠ CAUTION: Never fly your copter near people or over them. A propeller on a heavy multicopter, lifting a heavy camera, will work like a blender in your subject's face in the event of a likely mishap. Also note that in many areas it's illegal to fly model aircraft outside restricted areas.

K

with the equipment — not filmed at high speed and slowed down for a smooth look, and not edited in short clips, or stabilized in post-production.

I don't recommend 2-axis gyro gimbals. In my experience they introduce more shaking than they do good, even the very expensive ones. (And 3-axis gimbals introduce even more.) Since multicopters are extremely steady when it comes to holding direction, I don't think these are of any benefit.

Your best mount is something simple like a flexible plastic tube or soft foam. Just accept that the camera is not level at all times.

» Cameras and Video Downlinking
You *can* get really cool videos and pictures from multicopters if you've practiced flying, and if you use the medium on its own terms: accept the ever-moving picture, use a lightweight camera, and focus on action shots where the camera is moving through the air.

The best videos I've seen are using extreme wide angle, usually made with the GoPro camera brand (Figure K), which can also shoot at 60 frames per second (fps), giving a slow-motion feeling. The lighter the camera, the better the flight performance. Think 8oz and below.

Finally, your best tool is video feedback. Actually seeing what you film, while you're doing it, is called *first-person video* (FPV). There are many options for wireless video downlinks, depending on the following parameters:
» Cost, weight, and power consumption.
» How large an antenna can you carry to the field?

» What RF bands are allowed in your country?

» Which are already used on your copter?

» Transmitting power: systems 1 watt or stronger may require a ham operator's license. Frequency regulation information is available at makezine.com/go/hamradio.

» Electromagnetic pulses: powerful transmitters can make servos and other electronics malfunction. These things have to be experienced; there are no golden rules that I'm aware of. Sometimes things just interfere.

In general you're looking for lighter weight, longer range, less power consumption, and undisturbed frequencies. You can't expect to use cheap, random TV transmission gear. Get something from a shop that has experience with video downlinks from multicopters.

And if you get a pair of video glasses for monitoring (Figure L), you can see what the camera in the sky sees, even in sunlight. If nothing else, it's really cool to be able to elevate your field of vision by remote control.

Going Further: Drone Multicopters

Once you've mastered R/C multicopters, you might want to try drone multicopters.

When most people say *drone* they're talking about flying by GPS coordinates and waypoints in fully autonomous mode, and that's something special. One example is the ArduCopter, controlled by an Arduino-based autopilot developed by DIY Drones (Figure M).

There are also popular setups where cameras film the drone, and a computer calculates its flight from what the cameras see (little dots on the copter). Perhaps you could even set up a drone to navigate by the sun. It's all just sensors.

If you do experiment with drones, never let your autonomous machine go beyond visual contact. Most systems I know of have a built-in maximum range of 250 meters.

Once you start playing with multicopters, you'll notice there's no longer a sharp border between "autonomous" or "R/C" flight. Any multicopter is a robot that to some degree is autonomously controlling its motors (or it would crash). And even fully autonomous

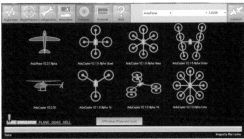

drones have the option of killing the automation and returning to R/C control (anything else would be hazardous).

With multicopters, it's always some form of R/C, and it's always some kind of autonomous. ▨

➕ More DIY multicopters and kits: scoutuav.com, multiwiicopter.com, kkmulticopter.kr

▣ Quadcopter FPV: makezine.com/go/fpv; 3D-printable quad: makezine.com/go/hugin

Frits Lyneborg runs letsmakerobots.com, the largest online community of its kind, which he started in 2008 as a forum for robot electronics, programming, funny ideas, and inspiration.

Max Levine (M. top); Michael Oborne (M. bottom)

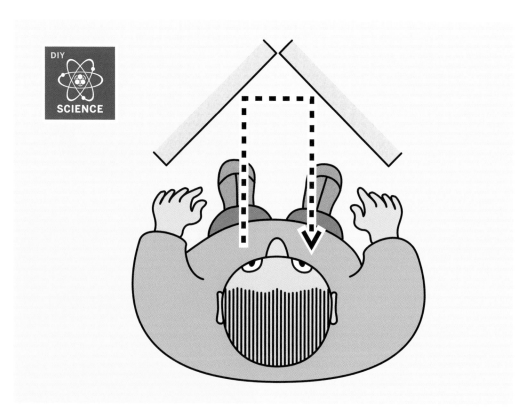

Flip Your Face

See yourself as others see you with a true mirror.

By Charles Platt

MOST PEOPLE BELIEVE THAT A MIRROR rotates your image from left to right, but this is not really correct. To clarify this confusing issue, I've sketched the story of an intrepid investigator named Ray, who reflects on it in some depth (Figure A). Ray happens to have a right ear that's bigger than his left, which makes it easy to see whether his image has been flipped. The lesson from this story is that when a photograph is turned around from the position in which it was taken, the image turns with it. Therefore, whether it looks as if your face is flipped depends on your point of view.

Can we create a mirror image that looks the same as a photographic image, so you see yourself as others see you? Two mirrors oriented at 90° to each other can produce this effect. Light from your face bounces off the first mirror and then the second mirror, rotating a total of 180° before it returns to your eyes. This combination of mirrors is often known as a "true mirror," although I think of it as an "untrue mirror" because of the rotation that occurs.

Whatever you choose to call it, building it is easy enough. You just need a pair of small, rectangular, unframed mirrors, some plywood, and enough glue and screws to hold everything together. The only challenge is to mount the mirrors very precisely.

1. Choose your mirror.

The best kind of mirror for this project is one with its reflective surface on the front rather than the back. This will almost eliminate the visible gap between the reflections where the mirrors meet. Ordinary back-surfaced mirrors

ADVENTURES IN ASYMMETRY   with "Reflective" Ray

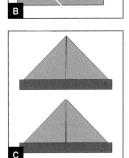

MATERIALS

Mirrors, thin, unframed, 6"×6" or larger (2)
Must not be beveled. For best results, use front-surfaced mirrors (aka first surface mirrors) such as those at surplusshed.com.
Plywood, ½" thick, 24"×24"
Wooden strips, ½" wide, as long as the width of each mirror (2)
Wood screws, flat-head, #6, ¾" long (14)
Scrap wood, small pieces for wedges
Construction paper, black, letter-sized (2 sheets)
Epoxy glue, 10-minute
Plywood or plastic (optional) for an enclosure

TOOLS

Work gloves and eye protection
Saw
Straightedge
Drill and bits
Screwdriver
Glass cutter (if needed)

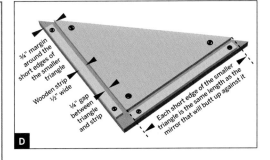

¾" margin around the short edges of the smaller triangle
Wooden strip ½" wide
¼" gap between triangle and strip
Each short edge of the smaller triangle is the same length as the mirror that will butt up against it

2. Create the mirror frame.

After choosing your mirrors, you need 2 wooden triangles that contain exact 90° angles. One way to obtain them is to buy a precut 24"×24" square of ½" plywood from a hardware store, and remove 2 corners with a saw, as shown in Figure B. Each short edge of each triangle should be the same length as the mirror that will be butting up against it.

Check that the triangles are accurate by standing them opposite each other on a flat surface, with their vertical edges touching. If there is the slightest gap at the top or bottom, one or both of their angles isn't exactly 90° (Figure C).

Now cut 2 larger 90° triangles from your same square of plywood, with their short edges 1½" longer than the short edges of the first 2 triangles. Screw each small triangle flat onto a large triangle, with the long edges flush

will work and save you money, but the reflected image will have a line down the middle.

Whichever kind of mirror you choose, you should wear work gloves while handling it. Eye protection is also advisable. Don't underestimate the danger of broken glass, which has scalpel-sharp edges.

The size of the mirrors is up to you, although if they measure less than 6"×6", your field of view will be limited. Some options are suggested in the parts list.

and a ¾" margin around the short edges. Add two ½"-wide wooden strips, leaving a ¼" gap, as shown in Figure D (preceding page).

Stand your mirrors vertically in the gaps, and insert 2 small wedges behind each mirror to squeeze it firmly against the smaller triangle, as shown in Figure E. Verify that the mirrors are precisely oriented, then mix some quick-setting epoxy, dribble it in behind the mirrors and around the wedges, and leave it to harden for at least an hour (Figure F).

Turn everything upside down, lower the mirror assembly into the gaps in the second set of triangles, and use glue and wedges as before. Your mirrors are now held securely between pieces of plywood above and below. Be careful not to touch the reflective surfaces during these procedures, because front-surfaced mirrors cannot be cleaned easily. To remove fingerprints, use Windex with an extremely light and gentle touch.

3. Finish and test.

Your true mirror will look nicer if you cover the interior plywood surfaces. I used some thin black felt that I happened to have, although construction paper would be just as good. I then mounted the mirrors in an enclosure, not just to protect them, but to protect people from the glass edges. Your design for a box is up to you. Personally I chose ABS plastic (Figure G), because it's so easy to work with (see makeprojects.com/project/a/1324).

Figure H shows some tests that you can perform, but here's the most important one. First, hold your true mirror facing you, with the joint between the mirror panels vertical. Check your reflection. Now keep the box facing you while you slowly rotate it till the joint between the mirrors is horizontal. You'll find that your image has turned upside down. Did you doubt me when I told you that a regular mirror doesn't rotate your image, but a true mirror does? Here you see the proof. ◪

E

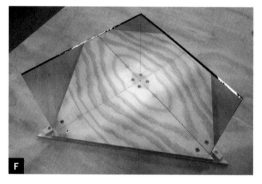

F

G

H

Charles Platt is the author of *Make: Electronics*, an introductory guide for all ages. A contributing editor of MAKE, he designs and builds medical equipment prototypes in Arizona.

Auditory Illusions

Explore your sense of hearing by creating weird sound tricks and effects.

By Michael Mauser

HAVING LONG BEEN INTRIGUED BY optical illusions, I was delighted to learn that there are also illusions we can experience with our other senses. Auditory illusions in particular fascinate me. Sound technicians and musicians often exploit these illusions, just as moviemakers and artists use visual illusions. Scientists also investigate auditory illusions to learn how we hear, and I've had a lot of fun duplicating their experiments and experiencing these tricks for myself.

To create the following 8 illusions, I used the free cross-platform audio editing software Audacity (audacity.sourceforge.net). Figure A (following page) shows Audacity's control panel and corresponding functions. I'll start with the simplest ones to familiarize

you with Audacity, and list each illusion with the approximate amount of time it takes to create. Many of these auditory illusions have visual counterparts, which you can see for yourself (Google them). It seems, the ears can fool the brain as easily as the eyes can.

Gregory Hayes

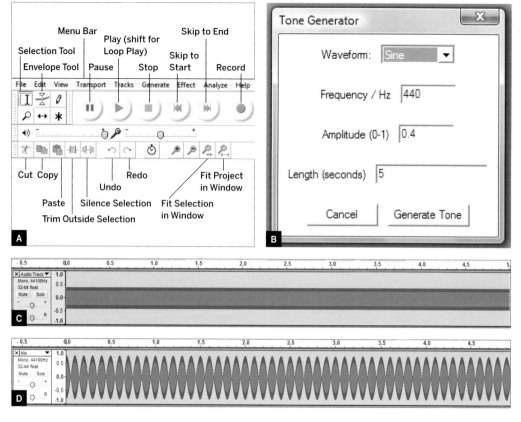

A

Menu Bar
Play (shift for
Loop Play)
Skip to End
Selection Tool
Skip to
Envelope Tool Pause Stop Start Record

File Edit View Transport Tracks Generate Effect Analyze Help

Cut Copy
Redo
Fit Project
in Window
Undo
Paste Silence Selection Fit Selection
in Window
Trim Outside Selection

B

Tone Generator

Waveform: Sine

Frequency / Hz 440

Amplitude (0-1) 0.4

Length (seconds) 5

Cancel Generate Tone

C

D

Binaural Beat (10 mins.)
"It only exists in your brain."

When 2 tones of slightly different frequencies play together, they create a *beat frequency* equal to the difference in their frequencies. This is because the 2 waves periodically align and then go out of phase, synchronizing additively and then opposing each other. When you tune a guitar, you listen for and eliminate the beat frequencies between strings.

When you use headphones to play a slightly different tone in each ear, the beat frequency between them doesn't exist anywhere, but your brain re-creates and perceives it anyway. To see how this works, here's how to create 440Hz and 450Hz tone tracks, play them together to generate a 10Hz beat, then play them separately in each ear and perceive a subtler version of the same beat frequency.

In Audacity, bring up the Tone Generator box (Figure B) by selecting Generate → Tone, then change the values to create a sine wave

at 440Hz, 0.4 amplitude, 5 seconds long, and click the Generate Tone button (Figure C). Hit the Play button to hear 5 glorious seconds of a pure "concert A" or A4, as musicians call this pitch. Click outside the track pane and create another tone track that's 450Hz but otherwise the same. Hit Play to hear both tracks, and the 10Hz beat.

Shift-click in the control area to select both tracks, and choose Tracks → Mix and Render (or Project → Quick Mix) to combine them. While the separate tracks showed smooth tones, the resulting combined track will show beats (Figure D). Play this track and it will sound the same as the 2 sounded together.

Back out with the Undo button, Ctrl+Z, or Edit → Undo, so the 2 mono tracks display again. Pull down the Audio Track Control Panel menu at the top left of each track pane and select Left Channel for one track and Right Channel for the other. Hold the Shift key and click the Play button; this will play the tracks in a loop until you hit Pause or Stop. Each tone now comes from a different speaker, so the beat you're hearing results from the 2 tones interacting in the air before before they reach your ears.

Now listen using stereo headphones. The faint beat you hear underneath the pure tones is something that your brain puts together after your ears have converted sounds into nerve signals. It can be subtle — try muting one track so you hear the difference between a pure tone and the mix with the binaural beat, then hold both headphones close to one ear to hear it monaural again. It's similar to the random dot stereogram visual illusion, in which your brain combines input from each eye into the perception of an image that doesn't exist.

Phantom Rotating Tone (10 mins.)
"It sounds like it's moving."
There are a few ways that our brains try to determine the location of things that we can hear but not see. A sound's volume in each ear usually corresponds to which ear its origin is closer to, but our brains also compare the phases of the sound waves as they reach each ear, and this can also contribute to our perception of volume. To imagine how this works, think of standing in the ocean up to your ears. If you're facing the waves as they come straight at you, each ear will experience the peaks and valleys of the waves at the same time. But if you turn so the waves come at you from one side, then the ear on that side will detect each peak a short time before the other ear does.

The Phantom Rotating Tone illusion comes from the same type of sound as the Binaural Beat, but uses 440Hz and 440.5Hz frequencies to generate a beat frequency of 0.5Hz. Create this illusion in Audacity by following the Binaural Beat instructions, but make the second track 440.5Hz rather than 450.

When you listen to these tracks with stereo headphones, your brain interprets the half-second phase difference between the tones as coming from a sound source that's rotat-

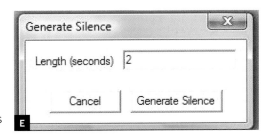

ing around your head. It will also probably seem like the sound volume is increasing and decreasing in alternate ears, but this is also an illusion; mute the sound to one ear and you'll immediately hear a steady tone.

Another way that we determine sound location (and a related illusion) is by detecting its mixture of frequencies. Our outer ears and heads block different frequencies in ways that we're accustomed to, which is why cupping your hands in front of your ears, which blocks the higher frequencies, makes sounds coming from in front of you seem like they're coming from behind you.

Some very realistic sound effects can be made by tinkering with the volume and phase of the frequencies that make up complex sounds, but we'll leave that to the big movie studios.

Franssen Effect (15 mins.)
"Sound from nowhere."
When a sound starts suddenly, we can often locate where it's coming from because the ear facing the sound hears it first. We can locate steady low-frequency sounds (like our 440Hz tone) because we detect a phase difference, as discussed above. And we can locate steady high-frequency sounds (3,000Hz or more) because they exhibit a shadow effect around the head, the way a post in the water will block short ripples more than longer waves.

But steady midrange sounds are harder to locate, especially when they reflect off walls. So when a 1,500Hz sound begins abruptly at one speaker, then fades there while simultaneously building at a second speaker, it fools our brains into thinking that it's still coming from the first speaker. This is known as the *Franssen effect* and it's similar to visual

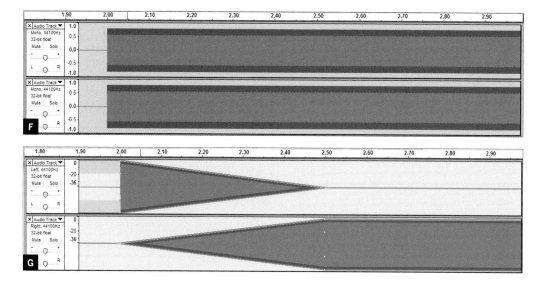

reflection illusions like Pepper's ghost.

Try this on friends in a room with normal reflective surfaces (e.g. no floor-to-ceiling curtains or tapestries). Use identical speakers about 10' apart, and stand centered about 10' in front of the speakers. You may have to unplug the first speaker to convince them the sound isn't coming from there!

In Audacity, select File → New, select Generate → Silence, and enter 2 in the box (Figure E, preceding page) to create a 2-second silent track. Play it; you'll hear nothing, but notice how the cursor sweeps from left to right across the track. To add a tone after the end of the silence, hit the Skip to End button (or select Edit → Move Cursor → to Track End). Now use Generate → Tone to add a sine wave at 1,500Hz, amplitude 0.8, length 28 seconds.

Select the track by clicking in the control area, then Copy, click outside, and Paste to make a second identical track. Zoom in on the first second of the tone by dragging the cursor over the track from a point just before it starts (2.00 seconds) to about 1 second later (3.00), and then clicking the Fit Selection button (Figure F). The 2.50-second mark should be somewhere in the middle. If not, just Undo, reselect, and Fit Selection again.

Use the track control menus as before to designate one track the Left channel and the other one Right. Select the Envelope Tool

icon in the top control bar and use it to fade one track out and the other one in between 2.00 and 2.50. You can do this by first clicking on the edge of a track to create "handles" (little white dots), and then using the tool to click and drag the handles up and down. Your screen should now look like Figure G.

To see the entire track, click the Fit Project in Window button. To play the illusion, click the Selection Tool, then Skip to Start, and Play. The sudden end of the tone at the end of the track will break the illusion by coming from what you thought was the inactive speaker, but you can fade it out and use Loop Play to make the illusion repeat without interruption.

Haas Effect (15 mins.)
"Sound that seems to vanish."

At times the signal from one ear can completely eliminate perception from the other. For echo delays of about 10 milliseconds (ms) our brains don't hear any delay — we just perceive a sound and its echo, reflecting from another direction, as a single sound coming from the original source. Sound engineers take advantage of this effect. By slightly delaying audio signal, they give an audience the impression that the sound is coming from a distant stage rather than nearby speakers. This is the *Haas effect*.

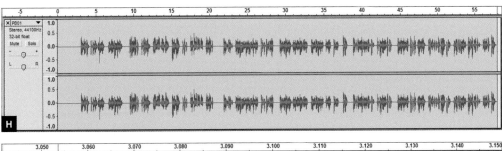

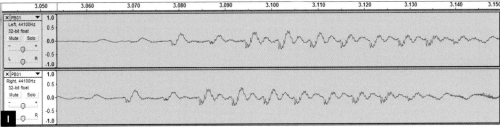

To experience the Haas effect, play a single-track voice recording on stereo speakers with a slight delay between speakers. In Audacity, start with a mono voice recording about 1 minute long. Record your own voice using the Record button, or else use File → Open to bring in a spoken-word MP3 file. With normal defaults, the recording will display as a stereo track (Figure H). To cut a longer track down to 1 minute, highlight sections and press the Delete key or click the Cut button. Click on Fit Project to expand the remaining audio track.

On the track's control menu, select Split Stereo Track and delete one of the 2 resulting tracks. Copy the remaining track to get 2 identical tracks, then change one track to Left and the other to Right, as before.

To add 10ms of silence to the start of one track, select the track, click Skip to Start, select Generate → Silence, and enter 0.01 for the seconds. Zoom in a few times by highlighting a short section of either track and clicking Fit Selection. You should see a 0.01-second offset in the tracks (Figure I).

Play these tracks on a stereo while you're between the 2 speakers, and you'll swear only one speaker is working! Your brain cancels out your perception of the delayed speaker, and just interprets it as the other speaker's being louder. (To do it with headphones, hold them in front of you, away from your ears.)

Verbal Transformation (10 mins.)
"Say what?"

Some words when repeated start to sound like different words, as our brains shift interpretation of where the words start and end. For example, *say* can be perceived as *ace*, and *rest* can be perceived as *tress* or *stress*. This audio illusion is called *verbal transformation*, and it's similar to the visual flip-flop you experience with an illusion like Necker's cube.

To experience verbal transformation using Audacity, record yourself saying "rest,"and trim the recording down to just the word. Loop-Play the word to hear the transformation. Try the same thing with "say" and "tress."

While you're recording, try a couple of other fun things. Record yourself reading a short phrase such as "free audio editing software" and then use Loop Play to hear it repeated. After a short while, it will sound like you're singing the phrase instead of reading it. Psychology professor Diana Deutsch reported on the first formal investigation of this illusion in 2008.

Now record yourself saying "Mom say yes" several times. Select the track and then choose Effect → Reverse from the menu. Play the result and you'll hear something like your voice saying "say yes Mom." Both *mom* and *say yes* are *audio palindromes*, a series of sounds that sound the same in reverse.

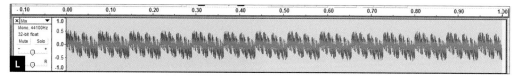

Note		A_0	A_1	A_2	A_3	A_4	A_5	A_6	A_7	A_8	A_9
						Frequencies used in making simple fractal sounds and their multiples.					
1x	13.75	27.5	55	110	220	**440**	880	1760	3,520	7,040	14,080
2x	27.5	55	110	220	**440**	880	1,760	3,520	7,040	14,080	28,160
4x	55	110	220	**440**	880	1,760	3,520	7,040	14,080	28,160	56,320
8x	110	220	**440**	880	1,760	3,520	7,040	14,080	28,160	56,320	112,640

K Frequencies used in making simple fractal sounds and their multiples.

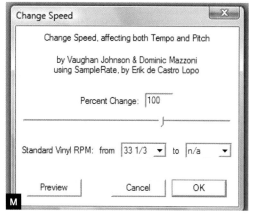

of 800Hz, and one track of 800Hz followed by 400Hz, also 0.25s each. Make one the left channel and the other the right (Figure J). The amplitude isn't important; I used 0.4.

Now Loop-Play the tracks through stereo headphones. Most people hear just one of the pure tones pulsing in each ear. Reversing the headphones should logically swap the tones between your ears, but for most people it doesn't — one ear seems to prefer hearing the high note while the other prefers the low.

Pull the headphones away from your ears and bring them out in front of you. You may find that you can maintain the illusion for a while before losing it, but once it's lost you have to bring the phones all the way back to your ears to recover it.

Deutsch compares the octave illusion to the Ames room and hollow face visual illusions; search YouTube to see how these work.

Deutsch's Octave Illusion (10 mins.)
"Flipping reality."

Diana Deutsch has investigated a number of audio illusions, including the octave (or two-tone) illusion. In this illusion, we hear 2 tones that are one octave apart (that is, they differ in frequency by a factor of 2) alternating between our ears. Both ears hear the same 2 tones for equal amounts of time, but our brains interpret it as one tone beeping in just one ear, alternating with the other tone beeping in the other ear.

In Audacity, use Generate → Tone to create one track of 0.25s of 400Hz followed by 0.25s

Fractal Sound (30 mins.)
"The same at twice the speed."

Most recorded sounds seem to have a higher pitch when played at double normal speed; recall the "chipmunk" effect with spoken word recordings. But German physicist Manfred

N	Frequencies used in making more complex ("descending") fractal sounds and their multiples.										
Note		F_0	$F_1\#$	G_2	$G_3\#$	A_4	$A_5\#$	B_6	C_7	$C_9\#$	D_{10}
1x	10	22	46	98	208	440	932	1,976	4,186	8,870	18,795
2x	21	44	92	196	415	880	1,865	3,951	8,372	17,740	37,589
4x	41	87	185	392	831	1,760	3,729	7,902	16,744	35,479	75,178
8x	82	175	370	784	1,661	3,520	7,459	15,804	33,488	70,959	150,356

Schroeder found that sounds with fractal waveforms (analogous to visual fractals, which exhibit the same pattern at any level of magnification) can sound the same or even actually seem to drop in pitch when played back at double speed.

It's easy to make a fractal sound that sounds the same when played back at double speed. As a nod to musicians, we'll make the pure tones one octave apart and centered around concert pitch (A4 = 440Hz), although other frequency spreads will also work. The table in Figure K shows the frequencies we'll use, with frequencies outside the typical range of human hearing in gray. Notice that the same sequence of tones repeats with every doubling.

To make this sound in Audacity, generate 11 sine wave tone tracks, each having one of the frequencies shown in row 1x. Make the durations all 1 second and the amplitudes all just 0.09, so that even when the peaks of all the frequencies coincide, the total will be less than 1, the maximum volume. After every couple of new tracks you create, combine them all together using Tracks → Mix and Render (or Project → Quick Mix) like we did in the first project. Your final track should look like Figure L.

Listen to the track. Then select it, choose Effect → Change Speed and enter 100 in the popup (Figure M). Listen again, and it will sound like the same mixture of pitches. Toggle between Undo and Redo to switch between playing both. Try doubling the speed of a recording of your own voice for comparison.

To make a fractal sound that decreases in perceived pitch when played at 2x, start with a series of tones that are slightly more than an octave apart. To give another nod to musicians, let's use an interval of an octave plus a semitone. This frequency ratio is

$2^{13/12}$, or about 2.12. If we center this interval scheme around 440Hz again, we get the rounded values shown in Figure N.

The notes A_4, $G_4\#$, G_4, and $F_4\#$ are highlighted. These notes (and all other such diagonal sequences) correspond to part of a descending semitone scale. Successively doubling the speed of a fractal sound composed of the frequencies in row 1x will result in the perception of this scale being played. If you doubled the speed of a more fully rendered fractal wave like this 13 times, you would arrive back where you started.

You now have the knowledge needed to create unique audio experiences. Many other audio illusions have been identified, and new ones are still being discovered. Have fun! ◪

Resources

For another endless scale illusion, see makezine.com/v/31.

To explore the science behind visual and audio illusions, check out Michael Bach's website (michaelbach.de/ot), Diana Deutsch's site (deutsch.ucsd.edu), and Deutsch's CD recordings *Musical Illusions and Paradoxes* and *Phantom Words and Other Curiosities*, from Philomel Records (philomel.com).

Michael Mauser (michaelwmauser@gmail.com) is a retired engineer and science teacher. His interests in biology, physics, and psychology led him to develop numerous sensory perception activities for the Arizona Science Center, and to document them in an upcoming book. He currently volunteers in the Expedition Health lab of the Denver Museum of Nature and Science.

Within its steel frame and aluminum skin, our 15-foot backyard rocket encompasses microcontrollers and LEDs, pneumatics, vibration and sound effects, and the joy of making.

Rocket-Ship Treehouse

BY JON HOWELL AND JEREMY ELSON

■ **RARELY DOES BUILDING A TREEHOUSE** require welding, grinding, painting, riveting, bending, crimping, plumbing, brazing, laser cutting, sound design, printed circuit board fabrication, distributed network protocols, an embedded operating system, sewing, and even embroidery. Ours did: a backyard rocket-ship treehouse.

The Ravenna Ultra-Low-Altitude Vehicle (RULAV), named for our neighborhood in Seattle, Wash., is a hexagonal capsule 7½ feet high, atop a tripod of the same height, for an overall height of about 15 feet. The frame is welded mild steel with riveted aluminum skin and a hinged entry panel and window. A ladder made of steel cable runs from the ground to the rocket's floor. A rigid interior ladder lets kids climb up and peek out the top.

Inside the rocket are nearly 800 LEDs forming dozens of flashing numeric displays spread across 14 control panels, each with an acrylic face laser-cut and etched with labels such as "Lunar Distance" and "Hydraulic Pressure." Working buttons, knobs, and switches operate the rocket's software.

Underneath the capsule are 3 "thrusters" that shoot plumes of water and compressed air under the control of a pilot's joystick, simulating real positioning thrusters. Takeoff and docking sequences are augmented by a pneumatic paint shaker that simulates the vibration of a rocket engine (Figures A and B, page 146).

Sound effects complete the illusion, with a powered subwoofer that gives the rocket a satisfying rumble.

An Engineering Playground

"It's for Jon's son, Eliot," was our justification to friends and co-workers after describing the growing list of planned features. As our ambitions spiraled and months of construction stretched into 2 years, it became transparent that the treehouse was just as much an engineering playground for the adults, a place for us to share our joy of making and teach it to the kids. Now that the rocket is complete, it's a fun plaything, but the journey was even more rewarding than the result.

The rocket was conceived in 2008 after Eliot's mom suggested that Jon install a swing set under the trees in the backyard.

"A swing set? *Everybody* has a swing set."

Mom said, "Then what *are* you going to build? A rocket?!"

"Yes. Yes we are. That is exactly what we're going to build."

Chassis and Skin

We went to Boeing Surplus and brought home a few big aluminum sheets and some sturdy aluminum tube. Our first idea was to build a geodesic structure formed entirely by bending and riveting, but early prototypes wouldn't stand up, proving that we really didn't know much about mechanical engineering.

Finally, we realized weight wasn't a design

Ravenna Ultra-Low-Altitude Vehicle

A

constraint for a rocket that never leaves the ground; it would be just fine to use steel. A new design was born: a welded steel chassis skinned with aluminum (Figure C).

The main challenge in constructing the chassis was the many compound angles required where the steel members meet. We proofed the design with a series of paper prototypes and built simple jigs that made it possible to make compound cuts with a steel cutoff saw. Wood jigs also helped greatly with accurate assembly (Figure D).

Once the frame was underway, we built a floor and a window screen from expanded steel, steel handles for the hinged pieces, and even a stainless-steel rope ladder (Figures E, F, and G). Eliot wasn't old enough to weld, but he pulled a lot of rivets (Figure H)!

B

Pneumatics and Plumbing

Once the chassis was done, we gathered a few friends around for an old-fashioned rocket-raising. We also dug a trench across our back-yard and plumbed it with ½" copper pipe.

A big green handle lets air into the rocket with a satisfying whoosh, and a needlessly elaborate "distribution manifold" has dials that spin and twitch as compressed air is used (Figure I). Visit rocket.jonh.net for a schematic of the final plumbing system.

Water for the orientation thrusters is supplied from a jug (Figure J). To refill the jug, a water line runs through the trench from the garden spigot (Figure K), where an electric solenoid lets the pilot "refuel" by pushing a button in the cockpit.

Electronics and Programming

While Jon was building the structure, Jeremy was working on the electronics. The original goal was modest: fill the rocket's interior with dozens of flashing lights and randomly changing numbers.

We designed a circuit board that would

C

WELD IT

Welding seems intimidating, but making basic joints in mild steel is pretty easy, and it's enough for a treehouse. You'll need a flux-wire welder, a helmet, gloves, an angle grinder, a wire brush, and perhaps an abrasive cutoff saw. Low-end versions of these can be had for around $300.

Buy mild steel ¹⁄₁₆"-wall square tubing from your local steel yard. (You can reclaim scrap if you clean off paint and rust, but don't mess with galvanized, chrome, or stainless; they make toxic fumes.)

Practice making a simple bead on one piece, then butt-joining 2 flat pieces, then joining on a right angle. Your goal is to use the heat of the arc to melt both work pieces. Cut your practice pieces apart to see that you really melted the metal on both work pieces, so that they froze back into one continuous piece of steel.

RIVET IT

Once you've got a steel frame, riveting sheet steel or aluminum to it is easy. You'll need a pop-riveter ($6), an electric drill, and a ⅛" bit, plus a few more for the ones you'll break. Drill a hole, place a rivet, squeeze the riveter until it pops, and repeat.

BRAZE IT

Some of our pneumatic plumbing is soft copper tubing with compression fittings that require only a pair of wrenches to install (ask your plumbing store how).

But those fittings are pricy, so where possible, we used sweat-fit fittings on ½" copper water pipe.

To get the job done, you'll need about $40 worth of tools: a tubing cutter, a pro-pane torch, plumbing solder, flux paste, and a plumber's wire brush.

Brush the corrosion off the outside of the end of a pipe and the inside of the fitting, smear flux on both surfaces, and push them together. Heat the joint until the copper subtly changes color, then touch the solder to the joint; when the joint is hot enough, the solder will get sucked in. Practice on scrap, then cut the joint apart to ensure that the solder filled it.

D

E

F

G

J

H

I

K

light up 8-segment numeric LEDs. They're cheap, bright, and available in a wide variety of colors and sizes. Each board holds up to 8 digits. Each digit's 8 segments are attached to one output of a 74HC259 latch, a small memory device that can be programmed to start or stop current flowing through each of 8 outputs.

To turn an LED segment on or off, the processor specifies the desired LED digit number using 3 output pins, and the segment number using another 3. Three pins also specify the board number (each board has switches that set its ID). Another pin specifies the desired LED state (on or off), and a final "strobe" signal indicates that all the other signals are ready.

Two layers of 74HC138 demultiplexers feed the strobe only into the intended latch (Figure L), and a single segment turns on or off as a result. A ribbon cable distributes the 11 control signals from a single processor out to all 8 boards. This way, 512 LED segments can be controlled by a single processor.

An early prototype worked, but even with only 2 LED digits, it took a week of evenings to construct (Figure M). We had to carefully modify a prototyping board with a rotary tool, and solder in each component and wire connection. This manual labor was time-consuming, error-prone, and not very fun. If we built all the boards by hand, our goal of "dozens" of LEDs would likely remain out of reach.

Sane treehouse builders might decide to scale back their ambitions. We went the opposite direction: why not design our own printed circuit board (PCB) and have it fabricated in bulk? The only problem was, we hadn't done anything of the sort before — in fact, we'd only recently learned how to light up an LED.

The thriving DIY community came to the rescue. We learned basic PCB design from online tutorials (Instructables, Adafruit, Seattle Robotics Society, UK Electronics

ETCH IT

Though there is a learning curve, creating your own printed circuit boards (PCBs) is surprisingly affordable.

Design your PCB in CAD software such as CadSoft Eagle, popular among hobbyists, or an open source alternative such as KiCad or gEDA. As you add each component to your circuit's schematic, a corresponding component appears on a physical layout drawing. Position these on the physical layout to minimize the distance between connected components. Then run the "auto-router" that converts the connections you've specified into copper traces.

PCBs can be fabricated at home in just a few hours (see MAKE Volume 02, page 164). However, dangerous acids are required, and it's hard to drill all the holes precisely.

Professionally made PCBs are worth the wait: they're far more precise (allowing more compact designs), can have more than one layer, and are covered with "solder stop" (unsolderable lacquer) that makes correct soldering much easier.

Sending your board out for fabrication is almost as easy as sending a PDF to a print shop. The Gerber format is the industry standard. If you have a small board, batch fabrication services can aggregate your design with other hobbyists', amortizing the setup cost; expect to pay about $5 per square inch for 3 copies. (DorkbotPDX runs an excellent service that anyone can use.)

For larger sizes or quantities, it pays to go directly to a board fabricator. We love OurPCB; they charge $50 setup plus 15 cents per square inch. We bought 20 display PCBs for the rocket for about $150.

Club) and discussion boards (SparkFun, AVR Freaks). We scoured the web for the best deals on board manufacturing.

After a few weeks of learning CadSoft Eagle, we had a layout we liked: 6.25" by 2.8" boards, routed tightly enough that a row of 0.56" LEDs could be mounted without gaps (Figure N). We sent it out for fabrication and spent the next 2 weeks giddy with anticipation. There are a lot of good reasons to become a maker, but perhaps none better than the joy of moments like when our first boards arrived (Figure O).

The rocket's final design uses 12 boards controlled by 2 processors that coordinate over a TWI network (2-wire interface, compatible with I2C). Not everything worked as expected: it turns out that running data lines in parallel over long distances without sufficient ground paths is not a good idea. We

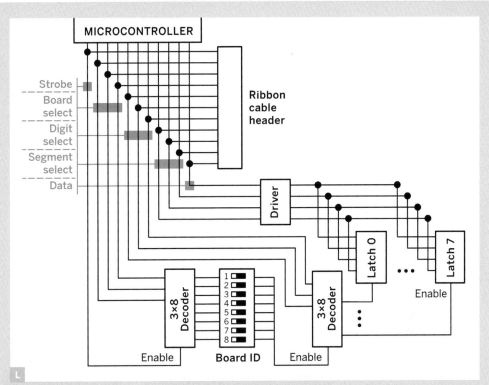

MICROCONTROLLER

Strobe
Board select
Digit select
Segment select
Data

Ribbon cable header

Driver

Latch 0 · · · Latch 7

Enable

3×8 Decoder

1
2
3
4
5
6
7
8

3×8 Decoder

Enable

Board ID

Enable

PROGRAM IT

It's never been easier to integrate microcontrollers into your DIY projects. Atmel's ATmega328 (the chip at the heart of the Arduino) has dozens of pins that you can use to drive logic, built-in processor-to-processor networking, a 6-channel analog-to-digital converter, a serial port, and much more. With 2K of RAM, it costs $4. For larger projects, the fancier ATmega1284 ($8) has 16K of RAM and more pins.

These chips come in easily solderable DIP packages and need no external components other than power/ground and a pull-up resistor on the reset pin.

Bring those 3 pins plus another 3 out to a standard 6-pin header, and your new circuit is programmable using any AVR in-circuit programming device, like Adafruit's $22 USBTinyISP. The software toolchain is free, including the compiler (GCC) and standard C library (*avr-libc*).

The Arduino is a great learning tool, but don't be afraid to add a bare microcontroller to your own custom board!

spent a few hours in the rocket with an oscilloscope (Figure P, preceding page).

Control Panels

We mounted the electronics in "control panels": aluminum boxes with transparent ⅛"-thick acrylic faces. We drew each face in Inkscape, placing rectangles where the circuit boards would go, and holes for standoffs and screws, then sent the drawings straight to a laser cutter to etch and cut the acrylic. (Laser cutters can be found at your local hackerspace, or try an online service.)

We made the pattern for each box by extending rectangles out from each edge of the face and adding ½" flanges all the way around. We printed the pattern and taped it to a sheet of aluminum, then marked the corners and holes with a punch, scored the lines with an awl, cut out the shape, and bent it on a homemade sheet metal brake. Each box is open at the bottom for cables to enter (Figures Q and R).

Standoffs attach the circuit boards to the acrylic face, and sheet metal screws attach the face to the box. The aluminum boxes are riveted to the rocket wall. The resulting control panels are sturdy enough to withstand a hapless foot.

Booster and Thrusters

Next we turned to the pneumatics that would power the "orientation thrusters" and "booster." The booster is a paint shaker that gives the rocket vibration during "takeoff." The thrusters are automotive engine-cleaning wands that aerate water using a supply of compressed air, producing a convincing jet blast of mist (Figure S).

Our original idea was to place air valves near the pilot's seat and direct air through hoses to the jets. A prototype didn't work very well — a hose itself can store enough air that the jets gave us unsatisfyingly soft whooshes

rather than the short, sharp blasts we wanted. Jon suggested putting the valves next to the jets, actuating them at a distance with bicycle cables leading down from the cockpit.

"Why not buy electrically actuated valves instead?" Jeremy asked.

"I don't know," Jon said skeptically. "Aren't they expensive?"

"I found some for only $15," Jeremy said. "Plus, the controller will be electrical rather than mechanical, meaning it becomes my problem."

"Sold!"

We bought a handful of solenoid valves from eBay (Figure T) and hand-built a simple electronics module that lets a processor control high-current devices. It has several power transistors, each with its gate connected to one of our boards' latch outputs. This lets software toggle the thrusters in exactly the same way as the LEDs.

Electrically actuated valves have another benefit: the pilot can use a joystick (Figure U). Vintage PC joysticks that use game ports rather than USB have simple electrical interfaces: 2 linear potentiometers, which vary between 0Ω and 100kΩ as the control stick is moved through the x and y axes. We connected those pots in series to fixed-value resistors, forming voltage dividers whose values are read by a processor's analog-to-digital converter. As the joystick is moved toward each corner of the rocket, the thruster in the opposite corner fires. The joystick's numeric x-y position is displayed on a panel, too.

Blastoff ... and Beyond

By now, the floodgates had opened. With all the lights and thrusters under software control, we realized the rocket's electronics could do far more interesting things than just display random numbers.

We created a takeoff sequence: once a launch code is entered into a 16-key keypad, a countdown is displayed on a control panel while audio from the real Apollo 11 sequence is played. At zero, the lights start to flicker, and the rocket starts to rumble from the movement of the paint-shaker and the bass

from our subwoofer.

In docking mode, the pilot guides a target onto a "docking clamp" represented by crosshairs drawn on a LED matrix display. Joystick motions produce action in both the real world (thruster firing) and the virtual one.

We even created a rocket version of the classic video game *Pong*, to keep crew morale high during long trips to the Moon.

By the project's completion, we'd written about 30,000 lines of C code, including a miniature operating system and a simulator that lets us test and debug the software on our desktop computers.

The completed rocket is quite a sight! At night, the glow of the LEDs is otherworldly,

and the illuminated water jets conjure dreams of space flight (Figure V).

Since Eliot was with us every step of the way, he also learned that toys aren't just something you buy, they're something we can build — together (Figure W). Because in the end, it's never really about the having. It's about the making. ◪

➕ For more photos, schematics, PCB layouts, and the continuing saga, visit rocket.jonh.net.

Jon Howell studies operating systems by day; after work he goes on adventures with his three junior astronauts. Jeremy Elson spends his free time flying airplanes, riding bicycles, and building electronics.

COUNTRY SCIENTIST
By Forrest M. Mims III,
Amateur Scientist

TRANSFORM THINGS INTO SOUNDS WITH THE PUNKPAC

Since Punk Science is the theme of this issue of MAKE, let's explore a new twist on electronic tone generators like the popular Atari Punk Console (APC). Why not hack the APC so that its tones can be controlled by light?

We'll do this by interfacing an optical fiber to a basic light-sensitive oscillator and an APC to create a Pixel-to-Audio Converter (PAC), which you could also call a Photon- or Photo-to-Audio Converter.

I call the resulting circuit a PunkPAC. It can be used to transform optical patterns formed by photos, fabrics, and even tree rings into variable-frequency tones for making music (or noise). It could easily be used for interactive concerts and museum exhibits.

Atari Punk Console

The 555 timer is the most popular of the many integrated circuits designed by legendary engineer Hans R. Camenzind. More than 30 years ago I described how to make a simple sound synthesizer by connecting the output of a 555 oscillator to the voltage control pin of a second 555 oscillator. This provided a stepped-tone generator in which slow pulses from the first oscillator altered an audio frequency tone from the second oscillator. The pulse repetition rate of each oscillator was controlled by a potentiometer, and appealing tone sequences could be created by adjusting either or both of them.

Electronic music experimenters began

experimenting with the circuit, which was dubbed the Atari Punk Console (APC) by Kaustic Machines (compiler.kaustic.net/machines/apc.html) because it makes sounds similar to the classic Atari 2600 video game console. The APC remains popular and even has its own Wikipedia page. Googling "Punk Console" yields more than 400,000 hits and over 1,000 video clips on YouTube.

Controlling the Punk Console with Light

The pots that control the frequency of a 555 tone generator and the APC can be replaced with light-sensitive photoresistors. This allows these circuits to be "played" by waving one's hands between the photoresistor(s) and a light source. Adding a DIY optical-fiber probe provides much finer control and the ability to transform natural and manmade patterns into distinctive tone arrangements.

→ START
Build an Optical Fiber Interface

A simple but sturdy optical PAC probe can be made from a jacketed plastic optical fiber (Jameco Electronics part #171272, jameco.com, or similar), a miniature CdS photoresistor with a dark resistance of 10MΩ or more (Jameco #202391 or similar), and a plastic applicator tip (#SIGSH10000 from hobby shops), as shown in Figure A. Follow these steps to make it:

1. The optical fiber comes in a 10' coil. Select the end with the cleanest cut. If both ends are rough, slice off one end with a sharp hobby knife. Make a second cut 10" up the fiber.

2. Use the hobby knife to slice through the long applicator tip 44mm from its small opening. The 10" optical fiber should fit through the hole in the tube. Recut if necessary.

3. Bend the leads of the photoresistor so they fit over the large end of the applicator.

4. Use long-nose pliers to form a small U shape in one end of each of a pair of connection wires that fit the holes in a solderless breadboard. Crimp a wire U around the bend in one of the photoresistor leads and solder

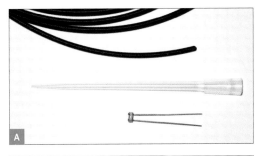

A

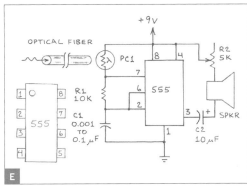

E

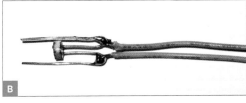

C

B

NOTE: Unless you plan to use the probe in a dark room, it's important to block all light from the back end of the probe.

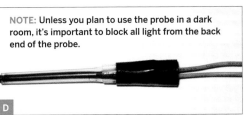

D

/ **Fig. A:** Components for a Pixel-to-Audio Converter (PAC) light probe.

/ **Fig. B:** Breadboard-style connection leads are soldered to the photoresistor leads.

/ **Fig. C:** The photoresistor is inserted into the large end of the applicator tip.

/ **Fig. D:** Assembled PAC light probe. The backside of the probe should be blocked from external light.

/ **Fig. E:** Basic 555 timer light-to-tone generator circuit.

it in place (Figure B). Repeat with the second connection wire and photoresistor lead.

5. Insert the photoresistor into the large end of the applicator (Figure C), and hold it in place while you push the optical fiber into the small end until it touches the photoresistor.

6. Wrap black electrical tape over the photoresistor leads to hold them in place and block stray light (Figure D). Or slide a 1½" section of black heat-shrink tubing over the photoresistor and warm it with a hair dryer.

PunkPAC: A Light-Controlled Tone Generator

Figure E shows a basic PunkPAC circuit made from a 555 tone generator. You can easily build it on a solderless breadboard. Insert the photoresistor leads from the PAC probe into the holes adjacent to pins 7 and 8 of the 555,

check your wiring, and connect a 9V battery. Switch off any lights, and the speaker should emit clicks or a tone. The frequency of the tone should rapidly increase when the optical fiber is pointed toward light. Increase the value of C1 to reduce the tone frequency. Try listening to the tone patterns and sequences that are produced when the fiber-optic probe is scanned across these words on a printed page or a computer monitor.

We'll return to more applications shortly, but first let's build an Atari Punk Console.

A Stepped-Tone (APC) PunkPAC

Figure F (following page) shows an Atari Punk Console circuit with a PAC probe. You can build the circuit on a solderless breadboard or assemble it from a kit version. Either way, carefully check the wiring before connecting a 9V battery.

The unique stepped-tone output from the APC differentiates it from the basic 555 tone generator in Figure E. Instead of a continuously variable tone controlled by light entering the PAC probe, the APC tone changes in distinct steps. The values for C1 and C2 in Figure F are very flexible. Experiment with different values to arrive at the best effects.

Applications for PunkPAC Circuits

Years is a clever creation by multimedia artist Bartholomäus Traubeck in which a record player spins a thin cross-section from a tree (traubeck.com/years). A light sensor detects changes in the annual rings and feeds the data to a computer that responds with piano notes. Since the rings aren't perfectly concentric, the sensor doesn't track individual rings but moves back and forth across them. The best musical bursts occur when it sees blemishes caused by age or disease (Figure G).

PunkPACs allow you to manually transform tree rings into variable or stepped audio tones more faithfully than the *Years* system. Illuminate a smooth cross section with a lamp and place the PAC probe over the wood (Figure H). Not much happens when you trace an individual ring. There's much more tone variation when you slide the probe across the rings to create an undulating sequence. Light-tinted early wood creates a higher frequency than darker late wood, so high-contrast rings work well. You can "play" a wide ring from a wet year, a thin drought ring, or the ring that grew the year you were born.

For better results, scan the probe across images of tree rings on a smartphone, tablet, or computer screen. For best results, convert them to high-contrast black and white, as in Figure I. For variety, try images of feathers, rocks, bark, fabric, and wallpaper.

Going Further

The PunkPAC lets you extract distinctive musical tones from everyday things, and its square-wave output can be processed by a microcontroller to trigger programmed notes emitted by various musical instruments.

Imagine a concert where you e-mail images or videos to the conductor, who scans them with a PunkPAC probe to embellish the music. Mine would be a video of sunrise in Hawaii (see youtube.com/fmims for a sample). ▣

Forrest M. Mims III (forrestmims.org), an amateur scientist and Rolex Award winner, was named by *Discover* magazine as one of the "50 Best Brains in Science." His books have sold more than 7 million copies.

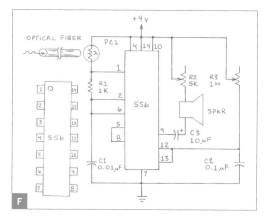

⚡ **Fig. F:** PunkPAC 556 stepped-tone circuit for use with a PAC light probe.

⚡ **Fig. G:** Bartholomäus Traubek's artwork *Years*.

⚡ **Fig. H:** Using the PAC light probe to scan the rings of a bois d'arc tree felled by a flood.

⚡ **Fig. I:** Using the PAC light probe to scan a high-contrast image of tree rings on a smartphone display.

Bartholomäus Traubeck (G)

1+2+3 Trash Can Composter

BY THOMAS J. AREY

I LIKE TO PICK THINGS OUT OF TRASH cans and reuse castaway items. Here, I repurpose the trash can itself to facilitate recycling organic waste into beneficial compost.

Commercial composting canisters can be costly, but they're simply a place to allow natural microbial processes to convert waste matter into a dark, fresh-smelling soil. Commercial versions allow air and some water to get in, and sometimes a way to mix.

Most home trash cans fail when part of the bottom wears away, leaving a hole. Such a trash can is perfect for this project, since we're just going to add more holes anyway.

1. Clean the trash can.

Scrub the trash can thoroughly inside and out to ensure that no inorganic waste remains. If you're squeamish about this, buy a new one.

It's helpful but not necessary to have a lid. If the original lid has gone missing, I'll leave it to you to come up with another solution.

2. Drill air holes.

Use a drill with a 1" spade bit to make air holes. Space the holes about 3" to 4" apart over all sides of the trash can. Drill plenty of holes, but don't compromise the structural integrity of the trash can. Avoid the corners to maintain the trash can's strength. Drill holes on the bottom to help with drainage.

3. Start composting.

Theories about composting are as numerous as the holes you have drilled. General rules:

» Keep the compost material damp, not wet.
» Mix brown material (such as leaves) in with green material (such as grass clippings).
» Use uncooked food scraps; no meat.
» Don't allow pet waste or anything treated with pesticides into your composter.
» Turn the compost material regularly. If your can has a tight-fitting lid, you can lay it on its side and roll it around on the ground. ◪

YOU WILL NEED

Trash can with lid
Drill with 1" spade bit

◪ **TIP:** Nothing goes to waste in this project! The 1" cutouts made by the spade bit can be used as insulating washers in other projects.

T.J. "Skip" Arey has been a freelance writer to the radio/electronics hobby world for over 25 years and is the author of *Radio Monitoring: A How To Guide.*

Damien Scogin

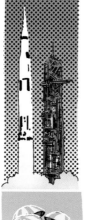

HOWTOONS.COM

SMALL STEP FOR MAN

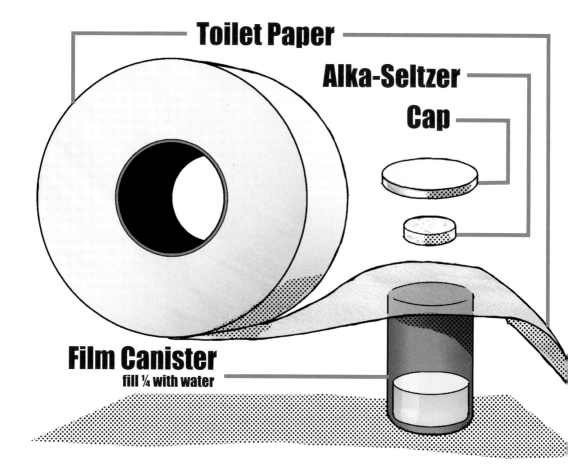

Toilet Paper

Alka-Seltzer

Cap

Film Canister
fill ¼ with water

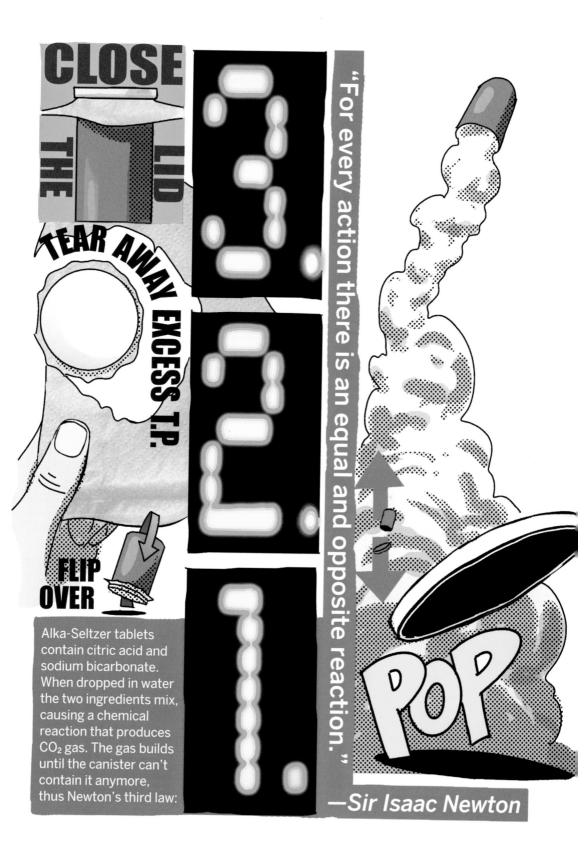

CLOSE THE LID

TEAR AWAY EXCESS T.P.

FLIP OVER

Alka-Seltzer tablets contain citric acid and sodium bicarbonate. When dropped in water the two ingredients mix, causing a chemical reaction that produces CO_2 gas. The gas builds until the canister can't contain it anymore, thus Newton's third law:

"For every action there is an equal and opposite reaction."

—Sir Isaac Newton

POP

**ELECTRONICS:
FUN AND FUNDAMENTALS**

By Charles Platt,
Author of *Make: Electronics*

THE CHING THING

You needn't believe in divination to enjoy making this fortune-teller.

In my previous column I described an electronic version of that old fortune-telling favorite, the Magic 8 Ball (*MAKE Volume 30, "Magic 8 Box"*). Now I want to delve deeper into divination — and the deepest you can go is surely the *I Ching*. This tool of prophesy originated in China thousands of years ago.

The *I Ching* supposedly offers insight into your current situation, and how it may change in the near future. You use a random system to draw 2 "hexagrams," each of which consists of 6 lines that can be solid or broken in the middle. (A solid line represents *yang* energy, while a broken one is *yin*, in case you were wondering.) The left hexagram interprets your curret situation. Some believe the right one suggests your future (Figure A).

To find out what the hexagrams mean, you look them up in a book. Dozens are available, but the one I use is *The I Ching: The Book of Answers* by Wu Wei, because its interpretations are dumbed down to a level where people such as myself, who are philosophically impaired, can make sense out of them.

You draw the hexagrams from the bottom up, by tossing coins — or by "casting yarrow stalks," if you're a traditionalist. (Yarrow is a weed, and you can find suppliers of dried stalks online.) Mathematically speaking, there are 4 possible present/future combinations for each pair of lines, as shown in Figure B.

Now, here's the catch. The combination probabilities in Figure B are what you get when you perform complex operations with yarrow stalks. Using coins will alter this pattern, giving every combination an equal probability. I wanted my electronic *I Ching* to be as authentic as possible, so I stuck with the ancient, yarrow-stalk probability set, even though this made the circuit a bit more complicated.

To display the hexagrams, I used light bars — little rectangles containing LEDs. Each line in a hexagram can be represented by 3 bars placed in a row, with the end bars lit and the center bar switched on to represent a solid line, or off to represent a broken line.

→ **START**
Choosing Chips

From Figure B, you'll see that the total chance of a line being solid on the left is 3 + 5 = 8 out of 16, and in 5 of those 8 instances, the line will also be solid in the hexagram on the right. As for the chance of a line being broken on the left, once again it is 8 out of 16, but the line will change to solid on the right in 1 of those 8 instances.

We need to make choices from 16 possible outputs, which prompted me to use the same 16-output 74HC4514 decoder chip as in the Magic 8 Box. In the box, I used rotational encoders to generate a random number. For my electronic *I Ching* (which I call the Ching Thing) the decoder is driven by a 74HC4520 divide-by-16 counter chip. By running the counter very, very fast and sampling it at an arbitrary moment, we can obtain a random number from its 16 possible states.

Now, how do we process this information to drive the light bars? Well, for one pair of bars, if the decoder delivers any of 8 numbers, we can switch on the center segment on the left-hand side. And out of those 8 states, 5 of them will be eligible to switch on the center segment on the right-hand side.

The easy way to arrange this is by using an OR gate with 8 inputs: the seldom-used but easily available 74HC4078. Check Figure C for a block diagram showing how this all works, and if logic gates are new to you, I hope you'll forgive me for mentioning that my own book, *Make: Electronics*, explains them in detail.

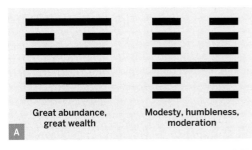

Great abundance,
great wealth

Modesty, humbleness,
moderation

Left (present) Hexagram	Right (future) Hexagram	Combination Probability
Solid line	Broken line	3 in 16
Solid line	Solid line	5 in 16
Broken line	Solid line	1 in 16
Broken line	Broken line	7 in 16

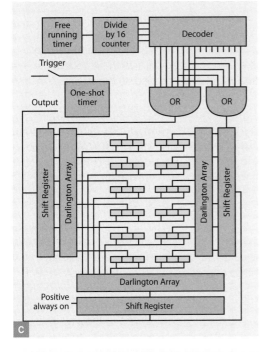

Fig. A: Two possible hexagrams describing your present and future. The meanings are quoted from *The I Ching: The Book of Answers* by Wu Wei.

Fig. B: For any pair of lines spanning present and future, there are 4 possible combinations, as shown here. The probability values are simplified if you toss coins instead of using the more traditional "yarrow stalks" to generate the lines.

Fig. C: Block diagram showing the basic components of the Ching Thing, and their functions.

Charles Platt

MATERIALS

Resistors, 5% tolerance or better: 10kΩ (1), 1MΩ (1), 200Ω (1), 1kΩ (2), and 100Ω (1)

Capacitors, ceramic: 1μF (1), 0.1μF (1), 0.33μF (1), and 0.01μF (3)

Capacitor, electrolytic, 470μF

IC chips: ALD7555 timers (2), 74HC4520 counter (1), 74HC4514 decoder (1), 74HC4078 OR gates (2), M74HC164 shift registers (3), ULN2003A Darlington arrays (3), and LM7805 voltage regulator (1) Part numbers may be preceded or followed by other letters identifying the manufacturer or package details; these letters can be ignored, so long as you're careful to buy through-hole chips (DIP or PDIP format), not surface-mount.

LED light bars (36) Lite-On model #LTL-2450Y

LED, 5mm for the prompt

Switch, SPST momentary pushbutton for the trigger

Switch, SPST toggle, 3A for a power switch

Switch, DPDT pushbutton for a reset switch that discharges the large capacitor

Power supply Any plug-in AC adapter with 9V DC output at a minimum of 1A. I used Triad Magnetics WDU9-1000.

Optional, recommended for testing:

Breadboards (2+)

Jumper wires

LEDs (30) Chicago 4302F1-5V (or F3 or F5), to show chip outputs

For the finished project:

Perf board, unplated

Hookup wire, solid core

Plywood or plastic sheet, ¼"

TOOLS

Soldering iron and solder

Wire cutters and strippers

To create broken lines in the hexagram on the left, we simply leave them switched off. So, the remaining 8 outputs from the decoder chip are unconnected — except that 1 time in those 8 instances, we need to light the center bar on the right. For this purpose, I connected one more output from the decoder to the right-hand OR gate.

Shift Registers

So far, I've described how to generate one line on the left and one line on the right. We need to do this 6 times to build complete hexagrams. Is there a way to memorize the first combination of lines, then repeat the process to create the next set of lines?

Yes, all we need is an 8-bit shift register. As

its name implies, it has enough memory locations for just 8 on-off bits of data. We load the first location with either a high state (on) or a low state (off), then send a clock pulse to the chip, which shifts the data to the next stage, so that we can reuse the first one. Another shift-register chip can control the right-hand hexagram, and a third chip will control the remaining light bars — the ones that are always on. This way, the bars of the hexagram will all scroll up together as the 3 shift registers move their data.

Our shift registers aren't powerful enough to drive all the light bars, so we have to amplify their outputs with Darlington arrays, which contain transistors to supply the necessary amperage. Their output side tolerates a higher voltage than the 5V DC required by logic chips. With a little trial and error, I found that if I connected all 4 LEDs inside a light bar in series by soldering pairs of pins together, (Figure D), I could run 9V through them, and they would take about 16mA, which is less than their rated 20mA. This eliminated the need for 144 load resistors!

Remember: each of the outputs of a Darlington array sinks current when it is "on." So you'll apply a positive voltage to each LED bar, and its negative end will be connected to each output pin of the Darlington array.

A 9V source can now run the whole circuit. But be sure to power the logic chips through a voltage regulator such as the LM7805, which lowers the 9V to the 5V DC that logic chips require. A 9V battery can't deliver the total peak current of 700mA that the circuit will draw, so you'll need an AC adapter rated for around 1A at 9V DC. You must add smoothing capacitors around the voltage regulator, to protect your logic chips from voltage spikes. See the schematic in Figure E. I added a simple power switch and a reset switch that discharges the large capacitor.

The Details

How will we sample the fast counter at random? By pressing a button. The same button will also tell the shift registers to shift their data along, ready to create new lines in the hexagrams, so you'll press the button 6 times to create your hexagrams.

Because buttons create noisy signals, I included a second timer (wired in one-shot mode) that transforms the button-press into a single, clean, one-second pulse.

Because we don't want the user to press the button too rapidly, I added an LED that acts as a prompt: it is illuminated when the system is ready, but goes off for 1 second after the button is pressed. Wait till the LED comes on before pressing the button again.

Note that the timers in this circuit are not old-fashioned TTL 555 timers. I have used a more modern CMOS version, because it has a higher output voltage, compatible with the HC family of chips in this circuit. Be sure to use the parts specified in the Materials list.

Test the circuit one chip at a time. Slow down the free-running timer with a 10μF capacitor and two 47K resistors, instead of the 0.01μF capacitor and 1K resistors in the schematic. You can monitor the outputs of the logic chips with little 5V DC LEDs such as the Chicago 4302F1-5V. These are handy things to have around, as they draw only 10mA, and their load resistors are built in. I spread my circuit across 3 breadboards (Figure F).

For the final version, each light bar measures about 0.2"×0.8", and each set of 3 can be glued into a slot in a box lid (Figure G).

Does the Ching Thing sound a bit … complicated? Almost like building a small computer? Indeed, but the chips are easy to wire together. I breadboarded the whole thing in a couple of hours. Also, the circuit can teach you some computer fundamentals. For instance, if you wonder why a component such as a shift register exists, consider that it takes a serial input and creates a parallel output — and if you don't know what a parallel port is, I'll leave you to find that out for yourself. You can ask the Ching Thing about it, although I have a feeling that Wikipedia may gave you a more helpful answer. Either way — may your fortunes all be electrically positive! ◪

➕ Get the *Make: Electronics* book from Maker Shed: makezine.com/go/makeelectronics

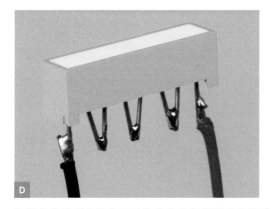

Fig. D: The 4 internal LEDs in a Lite-On LTL-2450Y light bar can be wired in series by bending the pins and soldering them together. Note the tiny notch in the plastic at the left end of the light bar, identifying its cathode (negative end).

Fig. E: Complete schematic for the Ching Thing.

Fig. F: The breadboarded version, using 5V LEDs to display the outputs and show the pin states of logic chips. The little LEDs can be a significant aid in debugging the circuit.

Fig. G: A possible plan for the lid of an enclosure for the finished version of the Ching Thing. White areas indicate cutouts.

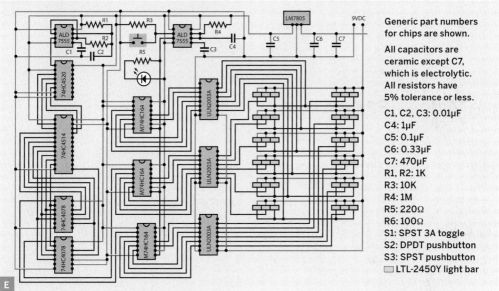

Generic part numbers for chips are shown.

All capacitors are ceramic except C7, which is electrolytic. All resistors have 5% tolerance or less.

C1, C2, C3: 0.01μF
C4: 1μF
C5: 0.1μF
C6: 0.33μF
C7: 470μF
R1, R2: 1K
R3: 10K
R4: 1M
R5: 220Ω
R6: 100Ω
S1: SPST 3A toggle
S2: DPDT pushbutton
S3: SPST pushbutton
▭ LTL-2450Y light bar

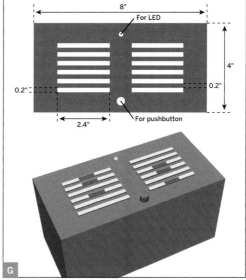

Charles Platt is the author of *Make: Electronics*, an introductory guide for all ages. A contributing editor of MAKE, he designs and builds medical equipment prototypes in Arizona.

A classic drill, the latest in nail guns, a mom-and-pop shop, and the synthesizer experience.

TOOLBOX

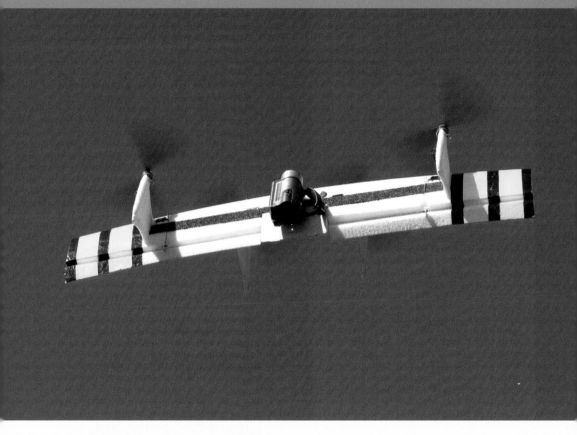

QUADSHOT MOCHA
$400 thequadshot.com

It's a quadcopter *and* a flying-wing — this unique R/C aircraft takes off vertically, then flies in copter or airplane mode, or both.

We received a pre-production Quadshot Mocha model, ready to fly. Like a UAV, it's got an autopilot board with an IMU that tells it where it's pointing and how fast it's rotating. Flight software lets you easily switch it to hover like a copter, fly easy like a trainer plane, or give you full control for aerobatics.

The hardware is open source and hugely hackable, with kits for DIYers. Deluxe models add magnetometer and barometer sensors, XBee radio, Gumstix WiFi/Bluetooth module, and CAN and I2C communication support.

Takeoff is copter-like; the 1-meter airframe is buffeted a bit, so it helps to goose the throttle for a bunny-hop. Flight is amazing, as the copter transitions seamlessly to airplane, and it's surprisingly forgiving — just center the stick to return to a reassuring hover.

—Keith Hammond

NWS Ergonomic Pliers

$35 makezine.com/go/ergopliers

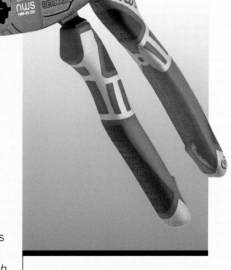

Ergonomic pistol-grip pliers offer a number of advantages over traditional-styled pliers. Whereas regular pliers point up at a 45° angle when held with a straight wrist, ergonomic pliers are bent forward such that the jaws are oriented in line with your arm.

As slight as this difference might seem, straightening your wrist when using pliers results in greater twisting and pulling power, improved reach in tight-quarters access, and reduced fatigue. Quality-wise, these pliers are absolutely fantastic. These are usually the first full-sized pliers I reach for when working inside a computer case or project box, where a high density of components and wires requires more controlled navigation.

—*Stuart Deutsch*

4093 Quad Oscillator Kit

$35 getlofi.com

If you've already built something like the Atari Punk Console (see page 152) and you're looking to expand your electronics knowledge, the Quad Oscillator is a terrific, inexpensive, and simple kit that makes a startling array of sinister noises. It comes with everything you need, but be sure to read the online instructions carefully, as the circuit board isn't labeled. Whether you need a little synth for your experimental noise band or just want to scare the pets, this is a fun and easy kit for a weekend afternoon.

—*Peter Bebergal*

ROCKIT SYNTH KIT

$149, $199 with case hackmeopen.com

For a more advanced project, The Rockit offers a true synthesizer experience, with MIDI jacks, a line in, and a nice array of knobs and switches.

This is an IC-heavy kit and requires careful and precise soldering. Some of the documentation is a bit unclear, but kit creator Matt Heins is quick to respond to emails, and he blogs frequently with updates and ideas.

Once complete, the Rockit can be used as a standalone instrument that produces lovely drones and loops. Tangerine Dream, eat your heart out.

—*PB*

ALLEN 31-PIECE MAGNETIC BALL END HEX KEYS

$53 www.allenhex.com

Allen's magnetic hex keys feature ball ends for improved access and straight ends for higher torque applications. Rare-earth magnets embedded in each ball end securely hold steel fasteners during positioning and install-ation. Compared to other fastener-gripping hex key designs I've used, Allen's magnetic keys are easier to insert into and release from fasteners.   —SD

Eureka! By Roy Doty
Escape!

DeWalt Nails It

DeWalt D51238K 18-Gauge 2" Brad Nailer Kit

$99 dewalt.com

Weighing in at 2.6 pounds, this brad nail gun feels solid, and the length of nails can range from ⅝" for delicate, shallow work, up to a maximum 2". I've been using the 2" nails with wood glue to help frame joints and corners before going back to install wood screws. Above the trigger is a rotating dial that adjusts how deep the nail will be driven in, any-where from flush with the wood surface to slightly raised.

I've been really impressed with the finish and quality of the tool and have not had any issues with jamming. Based on its performance and dashing good looks, I'd highly recommend this model to anyone looking for a new tool to help them with their woodworking projects this summer.

DeWalt D55153 Continuous 4-Gallon Electric Hand Carry Compressor

$250 dewalt.com

I can store this compact compressor under my workbench for quick access, and with the 4-gallon twin tanks it only takes 90 seconds to initially fill up. I can empty a clip of 60 brads at 80psi before the motor kicks on for another 25 seconds to restore pressure in the tanks.

The 120V 15-amp motor plugs into any home outlet. Each tank has a ball valve installed for draining moisture, so there's no fussing with threaded plugs or screws in hard-to-reach places, like other compressors. It's a great solution if you want to have compressed air in your garage for small tasks, such as using a brad nail gun, without investing in a large and expensive system.

—*Nick Raymond*

Wera Zyklop Ratchet

$60 www-us.wera.de

Wera's Zyklop ratchets have become my go-to choice for light- to medium-duty fastener-driving tasks. They feature locking swivel heads, 72-tooth gears (5° ratchet return angle), pushbutton socket quick-release, and exceptionally comfortable screwdriver-style handles.

Flip the driver into 180° inline mode, pop on a ¼" hex bit holder, grab onto the free-turning sleeve, and the Zyklop turns into a fast-action ratcheting screwdriver — one of the most versatile I've ever used. For higher torque, swivel the ratchet head back to the right-angle position.   —SD

Harry J. Epstein Co.

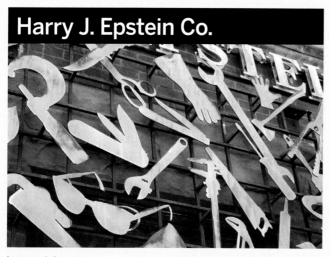

harryepstein.com

Harry Epstein is a small mom-and-pop tool distributor that specializes in American-made hand tools. Most of the products listed in their online catalog are new, some are new old stock, and some are surplus or factory seconds. Fantastic customer service, vast USA-made tool selection, and great prices; what more can one ask for? Box art? They'll do that too if you ask nicely. Browsing the actual store, located in downtown Kansas City, Mo., is said to be a wonderful experience.   —SD

GARRETT WADE YANKEE PUSH DRILL

$70 garrettwade.com

The first tool I used (for its intended purpose) was my father's Stanley No. 45 Yankee Drill. For those who don't know, a Yankee drill is a mechanical tool that's powered by you; it's simple, effective, and perfect for pilot holes, light drilling jobs, and tight spaces.

Around ten years ago, Stanley discontinued manufacture of the Yankee Drill. Fortunately, Garrett Wade stepped in to fill the void in the market. The GW version is every bit as hefty as the Stanley. It's a bit pricey new, but a Yankee drill is a great tool that I think every maker should have in their toolbox.
          —Michael Castor

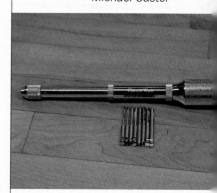

Natural Dyeing
Harvesting Color by Rebecca Burgess $23 Artisan

I recently experimented with using the oxalis plant (more commonly known as the weed sourgrass) to dye some towels a bright yellow. It was such an easy and rewarding process, I wanted to see what other colors might be within reach! This book has been a wonderful place to start. Author Rebecca Burgess does a great job explaining the nuts and bolts of the dying process, as well as the various properties of different plants that otherwise do the job of industrial dyeing chemicals. It's as much an exercise in chemistry and biology as it is a craft and cultural practice. Each plant and its corresponding color are beautifully presented with a map of its territory and season. Start hunting!

—*Meara O'Reilly*

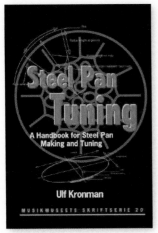

Steel Pan Alley
Steel Pan Tuning by Ulf Kronman Free hotpans.se/pan/tuning/index.php

When I was 16, I fell in love with the sound of the steel pan and decided that I was going to make my own. I did not produce one, but not for lack of trying.

Since my abortive attempts, much better information about the process has become available, and of course it turns out that it's a very demanding trade that takes years to master. If you want to learn, the best way is to serve an apprenticeship. On the other hand, if you don't have access to such training and still want to experiment with making your own steel drum (or just want to understand how it's done), you will not find a better reference than Ulf Kronman's book. Best of all? It's free.

—*Sean Michael Ragan*

New from MAKE and O'Reilly

LED Lighting: A Primer to Lighting the Future
by Sal Cangeloso $6 O'Reilly Media

Due to a combination of advances in technology, government legislation, and market forces, the LED lighting market is set to explode. This provides consumers (and, let's face it, geeks) with two interesting questions: what should we do now, and what do we need to know about the future? This book tackles both.

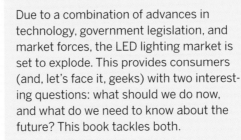

High-Tech Caveman
Fitness for Geeks by Bruce W. Perry $35 O'Reilly Media

I have friends who've been on the "caveman" diet for years (basically, eat as our caveman ancestors did), and I must say they're in fine fettle. I haven't been able to give up the odd donut, but I have been wanting to get more serious about my health.

Fitness for Geeks miraculously slid across my desk, and now I'm ready for action. In this "high-tech/ancient roots mashup," Bruce W. Perry lays out the current scientific thinking about diet and exercise (turns out those cavemen and women were pretty healthy); details modern fitness tools and apps like Fitbit tracker, Endomondo, and even the USDA National Nutrient Database; and dishes gym tips and lifestyle hacks.

Perry's book is refreshingly free of dogma; he simply shares everything he can find out about health with an audience who likes data. —Arwen O'Reilly Griffith

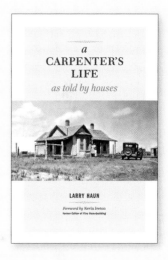

House Bound
A Carpenter's Life as Told by Houses by Larry Haun $23 Taunton Press

This is technically a memoir, but each passage in Larry Haun's life is told from the standpoint of a house and how it was built (many of which he was involved in). Making becomes metaphor in the most beautiful way: from a sod schoolhouse to Habitat for Humanity dwellings, from double-wide manufactured homes to adobes, this book is a poetic meditation on the way we live. While you probably couldn't build any of the houses simply from his descriptions, each chapter rings with the sound of nails in wood, from the mouth of someone who knows what he's talking about. Haun is a carpenter of the first order, and no shabby wordsmith, either.

—AOG

Getting Started with RepRap:
3D Printing on Your Desktop by Josef Prusa
$6 O'Reilly Media

This book provides an overview of RepRap, the open source 3D printer technology that's the basis of nearly every inexpensive 3D printer on the market. It teaches you what you need to know before you obtain a RepRap, and how to get the most out of it when you get one.

TOOLBOX

SILICONE GLUE BRUSH
$4 rockler.com

Apart from the small paddle at the other end of the handle, this is very similar to a chef's brush, although the bristles are a bit thicker and more widely spaced. If, like me, you use disposable foam brushes to save cleanup time, a silicone brush can be a more environmentally responsible, and, in the long run, cheaper option. You can use it like a disposable brush, except instead of throwing away the whole thing, you just throw away the blob of dried glue.

—SMR

SEYMOUR SUPER SHOVEL
$35–$50 seymourmfg.com

I got a shovel in high school, when I was first interested in gardening, and assumed I'd never need another. Now, that shovel is still perfectly functional and will always have a place in my heart, but I'm in love with my new best friend. The Super Shovel, aptly named, quite literally has teeth. Perfect for root pruning, it also makes digging through clay or rocky soil practically a pleasure. We recently put in some sidewalk gardens near our house, and this was the only shovel up to the task of loosening compacted city soil. I have the short D-handled shovel; it comes in a long-handled version as well.

—AOG

Parallel-Action Pliers

$45 grobetusa.com

A favorite tool of mine is parallel-action pliers. Difficult to find in retail stores, they can be located online and at some jewelry supply shops. The parallel jaw action gives a much larger and more even gripping area than standard pliers. Wire and rod pass through the pliers, which makes it easier to straighten sections of wire, make bends regardless of the position on the wire, and grip nails or wire close to where they penetrate a surface.

Depending on your desired use, you can find pliers with smooth or serrated jaws. Optional features available in various pliers include a spring-loaded handle, side cutters, and a wire groove in the jaws.

—Steve Crawford

Tricks of the Trade By Tim Lillis

Get the big picture.

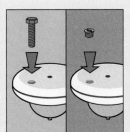

Looking to take a time-lapse panorama? Larry Towe of getawaymoments. com has a clever and inexpensive solution!

He used an Ikea kitchen timer, but this trick is adaptable to many timers. Drill a ¹⁵⁄₆₄" hole in the center of the top and insert a 1"-long ¼-20 setscrew.

On the bottom, drill an ¹¹⁄₃₂" hole and insert a ³⁄₈-16 to ¼-20 bushing. You may need to tap (add threads to) the hole with a ³⁄₈-16 bolt.

Now mount the camera and timer on a tripod (or add some non-slip tool mat to the bottom of the timer for tabletop use). Wind up and click!

Have a trick of the trade? Send it to tricks@makezine.com.

Polishing Pads

$50 makezine.com/go/polishingpads

Last spring I got it in my head to make a concrete bowl with broken-bottle glass aggregate. I tried every kind of sand-paper in the toolbox, to nil effect: scrubbing a quarter-sized area for half an hour wouldn't even begin to expose the glass aggregate beneath. I tried sanding wet and sanding dry. Then I found this set of eight "soft" polishing pads on Amazon, and took a chance. The business side of each 4"-diameter pad consists of a polymer honeycomb that looks sort of like the bottom of a sneaker. In use, the matrix slowly wears away, exposing fresh grit. Polishing the bowl was still a heckuva lot more work than I counted on, but I dunno how I would've done it at all without this set of pads.

—SMR

Peter Bebergal is the author of *Too Much to Dream*. He blogs at mystery theater.blogspot.com.

Michael Castor is the evangelist for the Maker Shed (makershed.com).

Steve Crawford works with Linux and PostgreSQL by day and sails, cycles, and tinkers as time allows.

Stuart Deutsch is a DIYer, and writes a whole lot more about tools at toolguyd.com.

Keith Hammond is projects editor at MAKE.

Tim Lillis is a freelance illustrator and DIYer (narwhalcreative.com).

Meara O'Reilly (mearaoreilly.com) is a sound designer.

Arwen O'Reilly Griffith is staff editor at MAKE.

Sean Michael Ragan comes from a long line of makers, and blogs at makezine.com.

Nick Raymond is intern coordinator of MAKE Labs.

✱ **Want more?** Check out our searchable online database of tips and tools at makezine.com/tnt.

Have a tool worth keeping in your toolbox? Let us know at toolbox@makezine.com.

✎ A German 50-pfennig note celebrates Otto von Guericke's famous demonstration of a vacuum — two teams of horses could not separate the hemispheres.

OTTO VON GUERICKE AND THE MAGDEBURG HEMISPHERES

Make an impressively effective low-tech vacuum pump.

In the spring of 1654, German scientist Otto von Guericke staged a dramatic demonstration of his new invention, the vacuum pump. Guericke was the mayor (*Bürgermeister*) of Magdeburg, and he arranged for a public demonstration in the town square. Two teams of 15 horses would try to pull apart two 20"-diameter metal hemispheres.

The hemispheres were not bolted, welded, glued, or otherwise mechanically connected to one another. Instead, the air inside the sphere had been evacuated by means of von Guericke's new vacuum pump. Strain and pull as they might, the force of the vacuum was stronger than the 30 horses — the teams could not part the sphere.

The Magdeburg Hemispheres, as they came to be known, were a pair of large copper hemispheres with mating rims. Von Guericke applied a thick coating of grease to the rims, attached his new vacuum pump apparatus, and pumped out a good bit of the air, producing a strong vacuum inside.

How hard would those horses have to pull in order to separate the hemispheres? To figure this out, first calculate the area of a 20"-diameter circle: $\pi r^2 = \pi 10^2 = 314$ square inches (in^2). Air pressure at sea level is about 14.7 pounds per square inch (psi, or lbf/in^2). If von Guericke's pump was capable of pulling a near-vacuum of about $1 lbf/in^2$, then the force required to pull the halves apart is $314 in^2 \times (14.7-1)\ lbf/in^2 = 4,300$ pounds of force.

The hemispheres could sustain a load of well over 2 tons!

→ START
Build Your Own Magdeburg Hemispheres

It's easy and fun to build your own Magdeburg Hemispheres, although we'll use aluminum cake pans instead of copper hemispheres. Hooking up horses to pull the pans apart is definitely optional!

1. Attach knobs to cake pans.
Drill a hole in the center of each cake pan the same diameter as the screws you'll use with the cabinet knobs (Figure A).

Place a dollop of silicone sealant on each hole (Figure B). Insert the screws through the holes and into the knobs, and tighten them (Figure C). The silicone will seal around the screws, making the connections airtight.

2. Make the vacuum release port.
Place one cake pan on a wooden block. Mark a point about 2" from the center knob and drill a ⅛" hole. Cover the hole with a piece of aluminized tape (Figure D).

3. Make the gasket.
Stack 8 sheets of newspaper, mark a circle

MATERIALS

Cake pans, 8" diameter (2) not nonstick
Cabinet knobs (2)
Wood screws, small (2) to fit knobs. The mounting screws that come with the knobs are probably too long.
Newspaper (8 sheets)
Water
Silicone cement
Cotton balls
Aluminized tape

Cigarette lighter fluid (naptha) or denatured alcohol

TOOLS

Drill and drill bits: 1/8", and one sized for your knob screws
Scissors
Marking pen
Nail, 6d
Match or lighter

slightly larger than the cake pan, and then cut out the paper circles (Figure E).

Cut a 4"-diameter doughnut hole in the center of the paper circles. Stack them neatly and soak them briefly in water. Remove them from the water as soon as the inside paper circles are wetted.

Lay the paper gasket across one cake pan as shown (Figure F).

4. Pull a vacuum.

Pour about 3 drops of fuel on a small cotton ball and place it near the center of the cake pan, but away from the vacuum release port (Figure G). Carefully light the cotton ball with a long match or fireplace lighter (Figure H).

Place the other cake pan atop the paper gasket, taking care to align the rims of the pans, one on top of the other.

The burning cotton ball produces water vapor. The vapor quickly condenses into liquid, producing a partial vacuum inside the container.

5. Bring on the horses.

Pull on the knobs to separate the 2 pans (Figure I). The halves will not separate, even if substantial force is applied!

6. Release the vacuum.

Use the nail to poke a hole in the aluminum tape covering the vacuum release port. Once the seal is broken, the 2 halves will release. ⊠

⚠ **CAUTION:** The burning cotton ball will locally heat the cake pans; avoid this area with your bare hands.

William Gurstelle is a contributing editor of MAKE. Visit williamgurstelle.com for more information on this and other maker-friendly projects.

William Gurstelle; Gregory Hayes (I)

Breadboard Bots

AT MAKER FAIRE I PICKED UP SAMPLES of an amazing material called Sugru. It's a patented silicone that's easy to use and makes soft, strong parts. Knead it like putty, then shape it. In 24 hours it permanently sets. It grips tight and holds onto almost any clean surface, but it stays flexible and pliant. Perfect for mocking up elastomeric (rubbery) parts. It's often used to effect quick repairs, reinforce stress relief on plugs and jacks, and make soft bumpers or grips.

I used it to create poseable electronic robots. The Sugru acts as a flexible joint and holds the parts together. Thread some stripped 22-gauge soft copper "bell wire" through short pieces of ⅛" plastic tubing and solder electronic parts to the wires. Add blobs of Sugru to make elbow, knee, and shoulder joints. You can leave the wire sticking out the bottoms of the feet for use with proto boards. Wire together the alligator-clip hands to grab and connect the bot to another wire, component, or bot.

This bot's photocell head is connected to a protoboard oscillator circuit through both his feet. When you shine a light on his face, the 555 timer circuit plays an audio tone. Sweet!

You can use Sugru to add all kinds of cool stuff to your bot besides electronic components: magnets, suction cups, Lego bricks, and more. Try it! Go online at makeprojects.com/v/31 for more info on Sugru and photos, circuits, and details on these Breadboard Bots. ◪

Bob Knetzger is an inventor/designer with 30 years' experience making all kinds of toys and other fun stuff.

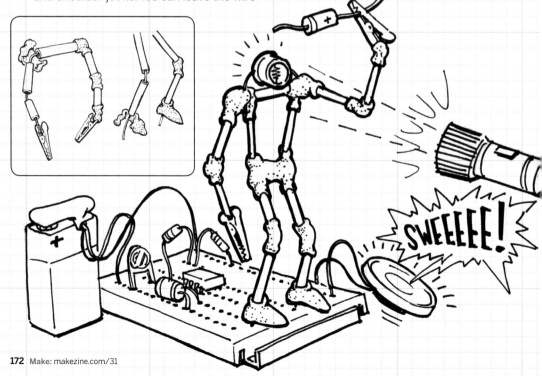

SWEEEEE!

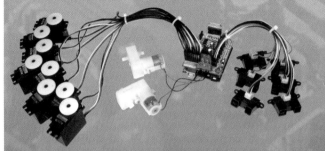

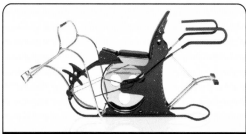
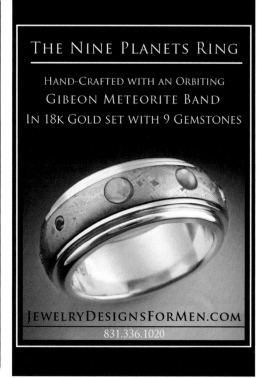

My 50-Acre Wi-Fi Network
BY KRIS KORTRIGHT

There are many aspects of life on a farm that can benefit from technology.

I STARTED A SERIES OF DIY PROJECTS
aimed at making life easier at Misty Brae Farm, a large horse farm and pony club riding center in Virginia. I have found that there are many aspects of life on a farm that can greatly benefit from technology. The farm owners had some specific requests, including a wi-fi network, video camera system, and riding lesson schedule management system. RiderNet is the first of these projects, which added wi-fi throughout the 50-acre farm.

To cover the entire farm, several of the wi-fi routers needed to be outdoors, mounted on trees and exposed to weather extremes. I looked at many enclosure systems and couldn't find the right one, so I decided to make my own using mineral oil as a medium. When submersed in mineral oil, the electronics work perfectly fine, and the oil provides thermal stability even when conditions are −50 or 150+ degrees (tested). I added a small pump and heater to keep the system above freezing in blizzards, and the custom waterproof acrylic tanks also prevent dust, bugs, and everything else from getting inside. I used acrylic for show, and because it's easy to see what's going on inside.

RiderNet is managed by an Arduino Mega, an Ethernet shield, and a custom RiderNet shield combo with a PowerSwitch Tail and five sensors, including internal/external temperature and a custom float sensor to detect the oil level. The wi-fi routers are linked together using XBees, Adafruit XBee adapters, and my XNP protocol, creating an out-of-band network that can be used to power-cycle wi-fi routers and get power/temp/oil sensor readings every 5 minutes from each unit. The management system was inspired by LadyAda's tutorials, and we plan to use RiderNet to carry data and instructions around the farm for all kinds of pony-related DIY projects! ▨

Kris Kortright is an internet technologist, DIY devotee, and horse enthusiast from Misty Brae Farm in Virginia.

Kris Kortright